AN APPROACH TO ARISTOTLE'S PHYSICS

AN APPROACH TO ARISTOTLE'S PHYSICS

With Particular Attention to the Role of His Manner of Writing

DAVID BOLOTIN

State University of New York Press

Published by
State University of New York Press, Albany

© 1998 State University of New York

All rights reserved

Production by Susan Geraghty
Marketing by Fran Keneston

Printed in the United States of America

No part of this book may be used or reproduced
in any manner whatsoever without written permission.
No part of this book may be stored in a retrieval
system or transmitted in any form or by any means
including electronic, electrostatic, magnetic tape,
mechanical, photocopying, recording, or otherwise
without the prior permission in writing of the publisher.

For information, address State University of New York
Press, State University Plaza, Albany, N.Y., 12246

Library of Congress Cataloging-in-Publication Data

Library of Congress Cataloging-in-Publication Data

Bolotin, David, 1944–
    An approach to Aristotle's physics : with particular attention to
the role of his manner of writing / David Bolotin.
        p.   cm.
    Includes bibliographical references and index.
    ISBN 0-7914-3551-2 (hardcover : alk. paper). — ISBN 0-7914-3552-0
(pbk.)
        1. Aristotle. Physics. 2. Physics—Early works to 1800.
3. Greek language—Style. 4. Philosophy, Ancient. I. Aristotle.
Physics. II. Title.
Q151.A8B65 1997
530—dc21   96-38143
                                                                           CIP

10 9 8 7 6 5 4 3 2 1

CONTENTS

Acknowledgments		*vii*
Introduction		1
Chapter 1	On the Principles of the Natural Beings	13
Chapter 2	The Question of Teleology	31
Chapter 3	On Continuity and Infinite Divisibility	53
Chapter 4	The Question of Place	77
Chapter 5	The Doctrine of Weight and Lightness	115
Chapter 6	On Aristotle's Manner of Writing	149
Index		*155*

ACKNOWLEDGMENTS

The bulk of this work was completed during the academic years 1991–94, while I was on an extended leave of absence from St. John's College. I am grateful to St. John's for granting me this leave of absence, and also to the Lynde and Harry Bradley Foundation and the National Endowment for the Humanities for the financial support that made it possible for me to accept it.

Chapter 2 is a slightly revised version of a lecture that I delivered at the Carl Friedrich von Siemens Stiftung in Munich on June 2, 1992. Chapter 3 is a slightly revised and shortened version of my previously published "Continuity and Infinite Divisibility in Aristotle's *Physics*," *Ancient Philosophy* 13 (1993): 323–40.

INTRODUCTION

Modern natural science emerged in the seventeenth century in explicit opposition to Aristotle's natural science, that branch of his philosophy that he called "physics." That the earth is not at rest in the center of the universe, but a mere satellite orbiting around the sun; that the stars and the planets are inanimate bodies made up of the same elements as bodies here on earth, and that their motions are subject to the same laws; that natural motion does not tend toward ends or fulfillments, but that every body in motion would continue indefinitely in a straight line if it were not for the action of external forces; all these and other such fundamental notions were regarded from the beginning as contradicting key doctrines of Aristotle's physics. The reigning belief in the truth of Aristotle's physics was the chief intellectual obstacle to the acceptance of the new science. And in the light of the success that this science has since enjoyed, it has generally been assumed that Aristotle's physics has been refuted, and that it is thus of little importance except as an object of historical inquiry.

Recently, however, a widespread and growing critique of modern science has led to a renewed interest in Aristotle's physics as something that can still teach us about the natural world.[1] This is not to say that any of the conclusions of modern science have been simply rejected in favor of the opposing Aristotelian claims. Even in the one most conspicuous instance in which a conclusion of modern science is rejected—namely, the Darwinian claim about evolution and its causes—those who deny that the origin of species can be explained on Darwinian principles alone do not propose a return to the Aristotelian doctrine that the species are eternal. And in any case, the rejection of the Darwinian account of evolution stems mostly from older and different sources than does the prevailing critique of modern science. More typically, we see an acceptance of the results of this science, combined, however, with a criticism of the ordinary interpretation of these

1

results; and this criticism has led to the view that modern and Aristotelian science, when both are rightly understood, are not opposed, but complementary to one another. The criticism to which I am referring begins by arguing that modern scientists themselves and other interpreters of their results are wrong to claim to have seen beneath the apparent world of our experience in the direction of its true or underlying being. For instance, Sir Arthur Eddington has frequently been taken to task for claiming that his solid and familiar writing table was largely illusory, and that the only table "really there" was a second, "scientific table," consisting "mostly of emptiness," through which "numerous electrical charges [are] running about with great speed."[2] The critics of such views acknowledge that modern science has succeeded in relating the particular events and particular objects of our experience to more general laws, laws that involve entities not accessible to direct experience. But it is denied that these laws or these indirectly accessible entities provide the answers as to what the world, or the events and objects of our primary experience, really are. To the argument that the beings as we think that we perceive them are merely relative to our own way of perceiving, it has been replied that the new entities disclosed by modern science are themselves relative to a sophisticated mathematical technique, and that this technique is dependent for whatever meaningfulness it has on a prior grasp of the world as it appears directly.[3] It is thus argued that only philosophy, as distinct from modern science, can tell us what the world most truly is. Now the philosophic inquiry that is called for by this critique is not traditional metaphysics: it is not an attempt to explain the world in terms of eternal and absolutely first causes.[4] Indeed, this critique renounces such an attempt as being incapable of success or even meaningless. What is called for, rather, apart from reflection on the acts of awareness, is a careful examination of the world as it appears with the aim of *describing* this world, or its primary phenomena, instead of "explaining" them in terms of anything else.[5] And Aristotle is turned to because his natural science has come to be seen as a masterful description along these lines. The traditional interpretation of Aristotle's science has been accused of concealing this fundamental feature of it, by misconstruing it as an attempt to derive the natural phenomena from ultimate causes.

And by contrast, one of the most thoughtful among the recent Aristotelian scholars has spoken of the *Physics* as a kind of phenomenology and as a study of the linguistic structures in terms of which we experience the natural world.⁶

There is, however, a great difficulty in reading Aristotle in this way. For although he is indeed a master at describing the world as it appears, and although he evidently sees this as one of his central tasks, he also evidently claims to be doing more than this as well. He explicitly denies, for instance, that the periphery and the center of the world, or the up and the down, exist only in relation to human beings; and he argues, on the basis of the motions we observe, that the earth is truly, as it appears to be, unmoved.⁷ Moreover, in his attempts to understand the perceived facts, he makes many assertions about what must exist and what must not exist even beyond the range of our perception. Thus, for instance, he does not limit himself to trying to speak about natural phenomena without positing the existence of an unperceived void, but he also argues that motion would be impossible through a void.⁸ And in keeping with this denial that there could be motion through a void, he claims that the motion of a projectile is not continuous, as it appears, but requires a series of "pushes" from the successively contiguous portions of the surrounding medium.⁹ Another example of a claim that goes beyond the perceptible realm is his argument that moveable bodies are not composed of (imperceptibly small) indivisible parts, but are infinitely divisible, i.e., divisible into parts that are smaller than any definite magnitude.¹⁰ And perhaps the most important example of a claim that transcends appearances is his teaching, already referred to, that the visible universe, including the species man, is eternal, and that its motions are all dependent upon the activity of a motionless Prime Mover.¹¹ Thus, Aristotle's natural science is not merely phenomenological, for it asserts, in various ways, that what lies beyond the range of perception is similar in character to what we do perceive; and it also asserts that this world is not merely an incidental effect of causes that might never have produced it, but a necessary and permanent product of the ultimate causes. But of course many, at least, of these assertions are difficult to swallow for someone who has been schooled in modern science. And so precisely those contemporaries who insist most strongly on our need to learn from Aristotle about nature have tended to distort him, if only by disregarding or playing down the importance of his claims that are at odds with modern science and by treating him merely as a student of the world as it appears.[12] And yet this approach leaves them open to legitimate criticism from the perspective of historical accuracy.

Accordingly, there is a certain case to be made for the more radical approach which acknowledges the importance of Aristotle's alien and speculative claims, but which nevertheless argues that modern science has not refuted him but has merely replaced his fundamental orientation with a new one. More generally, this approach argues that there can be no final truth, nor even progress toward such truth, in natural science, on the grounds that even what are called "facts" are experienced differently in the light of different overall perspectives, and that neither is there any other criterion that would allow us to justify one such perspective over the others.[13] And yet even this apparently more open-minded approach to the thought of the past implies a distinct rejection of Aristotle's view, inasmuch as he evidently thought that he was articulating the one true perspective on the natural world. And an interpreter who takes it for granted that Aristotle was wrong to understand his activity in this way will not have the incentive to study him with the kind of care that might disclose his deepest reasons for so understanding it. Thus, such an interpreter remains closed to the possibility of being instructed by Aristotle in case he was right.

And yet here we come back to the stumbling block of modern science. For the very resolve to consider it possible that some one perspective on nature may be the true one reminds us, in case we had forgotten, of the strength of its objections against Aristotle. Though there is surely much to question in the general approach of modern science, it still seems that at least some of its anti-Aristotelian claims are based on such strong evidence that we can not really doubt them. Thus, for instance, we can not seriously even entertain Aristotle's teaching that the moon and the other celestial bodies are imperishable, living beings.[14] We can not seriously regard the earth as being simply unmoved at the center of the world, nor can we believe that above the air there might be an element ("fire") that has no weight, but only lightness, or a tendency to rise to its proper place.¹⁵ We can not take seriously Aristotle's claim that a projectile is kept in motion by a series of "pushes" from successive portions of the medium.¹⁶ And whatever we may think of Darwin's account of evolution (i.e., as caused chiefly by random variation and natural selection), it is hard to treat Aristotle's claim that our own species has always existed as being even a respectable alternative. So precisely on the assumption that there is only one comprehensive truth about nature, it would seem that much, at any rate, of Aristotelian physics must now be rejected.

On the other hand, it is striking that many, if not all, of Aristotle's claims that are so at odds with modern science are at least equally at odds with most ancient science as well. Thus, for instance, there is no one among his predecessors in natural philosophy who is known to have held that the human race has always existed. His argument that there is a body with lightness, but no weight, is explicitly presented in opposition to the view of all his predecessors.¹⁷ And as for his teachings about the celestial bodies and about the motion of projectiles, only Plato among the philosophers gives any evidence of having thought along similar lines.¹⁸ So if Aristotle's positions on these and other such matters are mistaken, there was no need to wait for the emergence of modern science to become aware of the possibility that this is so. Moreover, once one begins to examine his supporting arguments with care, it becomes more and more apparent that he himself could not have accepted them all, and also that he did not in fact do so. It becomes more and more apparent, in other words, that he sometimes deliberately used inadequate or faulty arguments, and there even arise doubts as to whether, or in what sense, he accepted the conclusions they are meant to support.

Now admittedly, it seems strange at first even to entertain the thought that Aristotle might have misrepresented his views about nature. And yet considerable support for the suggestion that he did can be found among his ancient and medieval commentators. Themistius, for instance, writing in the fourth century of our era, practically begins his paraphrase of the *Posterior Analytics* by saying that "many of the books of Aristotle appear to have been contrived with a view to concealment."¹⁹ Simplicius, writing two centuries later, concludes the introductory section of his commentary on the *Physics* with a fuller statement of the same thought. He says that

> [Aristotle's] writings are divided into two classes, into the exoteric ones, such as his observational researches and his dialogues and in general those [writings] that are not concerned with the height of precision, and into the esoteric ones [or those that are meant for hearing only], which include this treatise [i.e., the *Physics*]; and in the esoteric ones, he deliberately introduced obscurity, repelling by this means those who are too easy-going, so that it might seem to them that they had not even been written.[20]

And still later, in the tenth century, the Islamic philosopher Alfarabi wrote in his *Harmonization of the Opinions of Plato and Aristotle* that

> whoever inquires into Aristotle's sciences, peruses his books, and takes pains with them will not miss the many modes of concealment, blinding and complicating in his approach, despite his apparent intention to explain and clarify.[21]

This consensus, then, across the centuries, that Aristotle disguised some of his thought should make it at least conceivable that he really did so. And thus, the mere fact that some of his apparently central teachings are now rejected with good reason by all competent judges does not prove that his own thinking even about these matters was mistaken.

Now at this point, the question arises of why Aristotle would not straightforwardly communicate whatever he thought about the natural world. I will try to respond to this question more fully at the conclusion of my study, after I present the evidence that he did disguise his thinking. For now, however, on the assumption that the evidence will prove persuasive, let me limit myself to two general remarks. First, one cannot begin to understand Aristotle's rhetorical posture without taking into account the *political* vulnerability of natural philosophy, and thus also of those who pursued it, in the ancient world. We are told, in fact, that late in his life Aristotle himself was formally charged with impiety and that he was compelled to flee from Athens in order to avoid the fate of Socrates.[22] And underlying this charge is the fact that the study of nature was not generally thought of as a respectable, or even a lawful, pursuit. Those who devoted themselves to natural philosophy were suspected of atheism, which was a serious crime; and the depth of this suspicion may help to make it intelligible that Plato's Socrates did not even attempt to dispel it at his trial, but rather suggested, falsely, that he had never been a natural philosopher himself.[23] Now it is conceivable that Aristotle could have lessened the threat of persecution by simply limiting the aim of his inquiry to a description of the natural world as it (generally) appears. But, to turn now to my second point, even apart from any curiosity about the ultimate origins of the appearances, he was compelled to make claims about these origins because he was aware—as, for the most part, both modern scientists and their phenomenological critics are not—that genuine science is impossible without at least some knowledge as to what they are.[24] Indeed, Aristotle's political situation, in which the study of nature itself was thought to be incompatible with the authoritative beliefs of the community, may even have made it easier for him to keep this fact in mind. And since he could not therefore ignore the question of ultimate origins, he thought it prudent, to the extent compatible with his primary aim as a teacher, to tailor his presentation of these and related matters so as to mitigate the hostility of the authorities.

The following studies will attempt to confirm this suggestion about Aristotle's manner of writing by examining his treatment of some important topics in the *Physics* and in *On the Heaven*. I will look, first, at his discussion of the principles of natural beings; then, at his teaching that there are final causes or purposes in nature; at his treatment of continuity in natural bodies and their motions; at his discussion of place; and, finally, at his account of the nature of light and of heavy bodies. With regard to these topics Aristotle is generally thought of, from the perspective of our modern knowledge, as having been naively speculative or even manifestly wrong. But I will try to show in all these cases not only that his genuine views are consistent with modern discoveries, but that they stem from a broader and deeper grasp than modern science possesses of the matters to which they are related. Now I have not even attempted here to discuss all the important topics regarding which modern knowledge, or modern prejudice, stands in the way of a genuine openness to Aristotle. In particular, I have not discussed, since I am not able to discuss adequately, the questions that surround his doctrine of the Prime Mover.[25] Still, I think that what I have done is sufficient in its breadth to make at least seem plausible what I believe to be the case—that Aristotle's physics is true in its fundamental claims, and that it surpasses all subsequent physics both in its articulation of the main features of the natural world and in its clarity about the central problems in the study of nature. And if I can make these assertions seem plausible, this work will have achieved its main goal, which is to encourage further study of Aristotle in order to help us learn what nature is.

NOTES

1. There has also been some interest arising from within modern science itself, as for instance in Heisenberg's appeal to what he calls the Aristotelian sense of "potentia" to characterize entities at the quantum level as they exist when they are not being measured or observed (*Physics and Reality* [New York: Harper and Row, 1958], 180–81). However, the appropriation of Aristotle's language to help make sense of quantum physics does not seem so likely to lead to a real openness to his central thoughts as does the critique of modern science that I discuss in the text.

2. Arthur Eddington, *The Nature of the Physical World* (New York: Macmillan, 1930), ix–xiv.

3. Thus Husserl argues, in *The Crisis of European Sciences*, that modern science was deceived into a "surreptitious substitution of the mathematically substructed world of idealities for the only real world, the one that is actually given through perception, that is ever experienced and experienceable—our everyday life-world" (Husserl, *The Crisis of European Sciences and Transcendental Phenomenology*, trans. David Carr [Evanston: Northwestern University Press, 1970], 48–49). Or, to take another formulation from the same work, Husserl claims that modern mathematical physics takes "for *true being* what is actually a *method*—a method which is designed for the purpose of progressively improving, *in infinitum*, through 'scientific' predictions, those rough predictions which are the only ones originally possible within the sphere of what is actually experienced and experienceable in the life-world" (51–52). Concerning the origin and character of modern mathematical concepts, see also Jacob Klein, *Greek Mathematical Thought and the*

Origin of Algebra, trans. Eva Brann (Cambridge: MIT Press, 1968).

4. Husserl does speak of pure consciousness as "absolute" being, in the sense that the existence of the stream of experience, precisely as what it presents itself as being, would not be called into question even by a denial of the (transcendent) existence of its objects. However, "absolute" being in this (Cartesian) sense does not present itself as the ground of its own existence, and thus Husserl acknowledges that "the transition to pure consciousness by the method of transcendental reduction leads necessarily to the question about the ground for the now-emerging factualness of the . . . constitutive consciousness." Such a ground, he continues, would "obviously transcend not merely the world but 'absolute' consciousness. It would therefore be an *'absolute' in the sense totally different from that in which consciousness is* an absolute, . . ." But he explicitly refuses to pursue the question of the existence of such an absolute. Husserl, *Ideas Pertaining to a Pure Phenomenology and to a Phenomenological Philosophy*, bk. 1, trans. F. Kersten (The Hague: Martinus Nijhoff, 1982), 134 (sec. 58; cf. secs. 44, 46, 49, 51n.).

5. For an early statement of the same, or a closely related, point of view, see Goethe, *Theory of Colors*, trans. Charles Eastlake (Cambridge: MIT Press, 1970): "We call these primordial phenomena, because nothing appreciable by the senses lies beyond them, on the contrary, they are perfectly fit to be considered as a fixed point to which we first ascended, step by step, and from which we may, in like manner, descend to the commonest case of every-day experience" (pt. 2, chap. 10, par. 175, p. 72). "But when even such a primordial phenomenon is arrived at, the evil still is that we refuse to recognize it as such, that we still aim at something beyond, although it would become us to confess that we are arrived at the limits of experimental knowledge. Let the observer of nature suffer the primordial phenomenon to remain undisturbed in its beauty; let the philosopher admit it into his department, and he will find that important elementary facts are a worthier basis for further operations than insulated cases, opinions, and hypotheses" (pt. 2, chap. 10, par. 177, p. 73).

6. Wolfgang Wieland, *Die aristotelische Physik: Untersuchungen über die Grundlegung der Naturwissenschaft und die sprachlichen Bedingungen der Prinzipienforschung bei Aristoteles*, 2nd ed. (Göttingen: Vandenhoeck and Ruprecht, 1970), 334, 338–40. Cf. Martha Nussbaum, "Saving Aristotle's Appearances," in *The Fragility of Goodness* (Cambridge: Cambridge University Press, 1986), 240–63.

7. *Physics* 205b31–34, 208b12–19; *On the Heaven* 294a10–297a6.

8. *Physics* 214b28–215b23.

9. *Pushes* is not precisely the right word for what Aristotle has in mind here, since his claim is that the portion of the medium in contact with the projectile causes it to continue its motion even when that portion of the medium is no longer in motion itself. *Physics* 266b27–267a18.

10. *Physics* 234b10–235b5, 237b9–22.

11. *Physics*, bk. Θ; *On the Heaven*, bk. A; *Generation of Animals* 731b18–732a1; *Meteorology* 339b28–30.

12. Wieland, for instance, acknowledges in his concluding remarks that the last book of the *Physics*, with its argument for the existence of an unmoved Prime Mover, goes beyond the scope of phenomenology. But he does not himself devote any thematic attention to this fact (Wieland, *Die aristotelische Physik*, 335–38).

13. See especially Heidegger, *What is a Thing?*, trans. W. B. Baron, Jr. and Vera Deutsch (Chicago: Regnery, 1967), 80–95. Also see Thomas Kuhn, *The Structure of Scientific Revolutions*, 2nd ed. (Chicago: University of Chicago Press, 1970), esp. 118–25, 205–7.

14. *On the Heaven* 283b26–284a14, 292a18–21.

15. *On the Heaven* 311a15–21, 311b13–312a8.

16. For an early, and convincing, refutation of Aristotle's claim, see Galileo, *Dialogue Concerning the Two Chief World Systems*, trans. Stillman Drake (Berkeley and Los Angeles: University of California Press, 1970), 149–54ff.

17. *On the Heaven* 308a7–29.

18. Plato, *Timaeus* 40b4–6, 79a5–80a2.

19. Themistius, *Analyticorum Posteriorum Paraphrasis*, in *Commentaria in Aristotelem Graeca* vol. 5, pt. 1, ed. M. Wallies (Berlin, 1900), 1. Except where I refer only to some particular translation, and not to the original text, all translations are my own.

20. Simplicius, *In Aristotelis Physicorum Libros Quattuor Priores Commentaria*, in *Commentaria in Aristotelem Graeca*, vol. 9, ed. H. Diels (Berlin, 1882), 8. Cf. Plutarch, *Life of Alexander* 7.5–9. And see also Descartes, *The Principles of Philosophy*, in *The Philosophical Works of Descartes*, trans. Elizabeth Haldane and G. R. T. Ross, vol. 1 (London: Cambridge University Press, 1972), 206: "The first and principal whose writings we possess, are Plato and Aristotle, between whom the only difference that exists is that the former, following the steps of his master Socrates, ingenuously confessed that he had never yet been able to discover anything for certain, and was content to set down the things that seemed to him to be probable, for this end adopting certain principles whereby he tried to account for other things. Aristotle, on the other hand, had less candour, and although he had been Plato's disciple

for twenty years, and possessed no other principles than his master's, he entirely changed the method of stating them, and proposed them as true and certain although there was no appearance of his ever having held them to be such."

21. Alfarabi, *Harmonization of the Opinions of Plato and Aristotle*, unpublished translation by M. Galston, p. 9. Quoted with permission of the translator.

22. See, for instance, Anton-Hermann Chroust, *Aristotle* (Notre Dame: University of Notre Dame Press, 1973), 1:145–54. For a broader picture of the political situation of philosophy in the ancient world, see Peter Ahrensdorf, "The Question of Historical Context and the Study of Plato," *Polity* 27 (1994): 113–35, and also Eudore Derenne, *Les Procès D'Impiété Intentés aux Philosophes à Athènes au Vme et au IVme Siècles avant J-C* (1930; reprint, New York: Arno Press, 1976). As for the connection between these political facts and Aristotle's rhetorical posture, consider the following statement by Hobbes: "And this shall suffice for an example of the Errors, which are brought into the Church, from the *Entities*, and *Essences* of Aristotle: which it may be he knew to be false Philosophy; but writ it as a thing consonant to, and corroborative of their Religion; and fearing the fate of Socrates" (*Leviathan*, ed. C. B. Macpherson [Harmondsworth: Penguin Books, 1968], end of fifteenth paragraph of chap. 46, p. 592).

23. Plato, *Apology of Socrates* 18b4–c3, 19b2–d7, 26d1–e2; contrast, however, *Phaedo* 96a6–99d2ff.

24. Cf. Leo Strauss, "On a New Interpretation of Plato's Political Philosophy," *Social Research* 13 (1946): 338–39.

25. These questions include even the one regarding projectile motion, for the following reason. In making his case for the existence of an unmoved Prime Mover, Aristotle's strongest argument against an infinite regress of motive causes (which would themselves be in motion) relies on the assumption that a motive cause must remain in contact with the body that it moves in order for the latter's motion to continue. For on this assumption (along with one or two others), Aristotle can argue that infinitely many moved movers would be like a single body moving with an infinitely great motion in the finite time of the original motion to be explained, which is impossible (compare *Physics* 256a13–19 with 242a49–243a31; see esp. 242b53–63; cf. 266b27–267a2). This assumption, in other words, is needed to exclude the alternative that a body's motion might be caused by a chain of successive interactions, stretching infinitely far back in time, among moved movers. And Aristotle's suggestion about projectile motion, according to which the various parts of the medium along the path of the projectile keep it moving, is more in

accord with this key assumption (or more nearly in accord with it) than is any account in which the body's motion could continue on its own (*Physics* 267a3–12; cf. n. 9). Thus, his rejection of the notion of inertial motion, or of any account that would allow a body to preserve an impressed motion, even for a while, on its own, is bound up with his argument for the existence of an unmoved Prime Mover (consider, however, *Physics* 254b33–256a3).

CHAPTER 1

On the Principles of the Natural Beings

In the first book of the *Physics*, Aristotle presents in outline his understanding of the principles of the natural beings. According to this account, natural beings come into being from form and from the underlying substrate or, in other words (since the substrate is itself twofold), from form, substrate, and the privation or lack which belongs to this latter when the form is not yet present.[1] Thus, for instance, to take an analogy from the arts, a bronze statue comes into being from its form as a statue, the bronze in which that form comes to be present, and the shapelessness of that bronze prior to its being made into a statue. Now after his initial account of these three principles, Aristotle goes on to say that only in this one way (μοναχῶς οὕτω, 191a23) can the perplexity of the ancients be resolved. The perplexity that he has in mind here had led some of the early philosophers to the paradoxical conclusion that nothing either comes into being or perishes. For those, he continues, who first, in accord with philosophy, sought the truth and the nature of the beings were misled by their inexperience into making this claim, on the grounds that what came into being would have to do so either from what is or from what is not, and that both of these are impossible. For as Aristotle has said in an earlier passage, all those who are concerned with nature agree in the opinion that nothing can come from nothing, or from what is not. And these early philosophers seem also to have held that there is no coming into being from what is, or even of what is, since what is, or that from which alone coming into being would be possible, already exists. By contrast, Aristotle says that his own account of the principles of the natural beings allows him to do justice to the appearance that there is coming into being. He explains that "we" (i.e., he and his school) also say that nothing comes into being simply from what is not, but (say) that this nevertheless happens, as by concomitance, since beings do come into being from privation, which is not (anything) in itself, but which exists as a concomitant in the substrate from which a being comes to be. Likewise, Aristotle continues, (we say that) there is no coming into being from what is, or of what is, except by concomitance. There is no coming into being simply of what is, his argument implies, since that from which anything comes to be already is, so that the thing does not come *into* being insofar as it is; and yet it remains true in a sense that what is comes into being, and that it does so from what is, inasmuch as one kind of being comes into being from another.² Thus, there is coming into being from what is, though not insofar as it is, and also from what is not, though only in the sense that the eventual form is not yet present.

This response to the philosophers' perplexity about coming into being would be a straightforward application of the principles which Aristotle has laid out were it not for his choice of a most peculiar example to illustrate one kind of being coming into being from another. For rather than saying, for instance, that a statue comes into being from bronze, he asks us to consider what it would mean if one kind of animal, a dog, were to come into being from another kind, a horse.³ This example is so bizarre that modern editors have been tempted to emend the text so as to make it read "if a dog were to come into being <from a dog or a horse> from a horse," and a number of English translators have translated the passage in this way. However, the surviving manuscripts are unanimous in support of the former, more difficult, reading.⁴ And this difficulty helps call our attention to another surprising feature of Aristotle's discussion. For after presenting the response that I have outlined here, he speaks of it as "one way" (εἷς μὲν δὴ τρόπος, 191b27; cf. 191a36) of responding to the perplexity of the ancients, and he follows it by mentioning another response (ἄλλος δ' 191b27). This other response, moreover, is based, not on his three principles, but rather on the distinction between potency and being at work, which he tells us has been elaborated more precisely elsewhere.⁵ And yet we recall that he had introduced the discussion by saying that it is "*only* in this *one* way" (emphasis mine), that is, on the basis of his three principles, that the perplexity of the ancients can be resolved. Now if we assume that our manuscripts are correct, and that Aristotle meant what he wrote, these difficulties, taken together, invite the suggestion that he himself may think that the first of his two responses merely *appears* to resolve the perplexity, and that the second one, which may not really even attempt to *resolve* it, is nevertheless a better or more truthful response.[6] This suggestion itself, of course, is in need of interpretation. But in order for me to explain it and to confirm that it is legitimate, we first need to look more closely at the principles on which Aristotle appears, at any rate, to rely in responding to the philosophers who denied becoming.

Aristotle's account of the principles of the natural beings takes its cue from the way we speak about coming into being in general. Our speech suggests, in the first place, that beings do come into being. For if the beings that we speak of according to their various forms—such as dogs, cats, or even statues—are truly beings, and not mere modifications of some other substance, then there are at least some beings that come into being. And a sign that a statue, for instance, is indeed a kind of being is that we do not say that bronze becomes "statuey," or even a statue, as we do say that a man becomes healthy or a general, but rather that *from* bronze there comes into being a statue. In other words, even though bronze persists as such in its transformation into a statue, we do not speak of being a statue as a mere modification of the bronze.[7] Our speech also suggests, in the second place, that there must always be some underlying thing from which a being comes to be, and that this substrate, though one in number, is more than one in kind. We say, for instance, that it is an unmusical (i.e., uncultured) human being who becomes a musical one; and his unmusicalness, which does not survive his becoming musical, is different in kind from his being human, which persists throughout the change. Now this illustration is not, to be sure, a case of the simple coming into being of a new being, but rather one of the qualified coming into being of a new attribute in a being that persists. But having begun from the way we speak in these more evident cases, Aristotle adds that it would become clear to one who reflects that even beings themselves always come into being from an underlying thing, as animals and plants do from a seed, and that this thing is both what it is as such and also something lacking in the eventual form. Thus, he says, it is clear, "if there are causes and principles of the natural beings, from which primarily they are and have come into being, not by concomitance, but each [as] what it is called according to its being [κατὰ τὴν οὐσίαν], that everything comes into being from the substrate and the form."[8] And he goes on to repeat that this substrate, though one in number, is both what it is as such and also that which contains, by concomitance, the privation of the eventual form.

Now Aristotle does not give a thematic account of the mode of being of these principles or of the way in which they are responsible for natural beings. But it appears at first, at any rate, that the principles are elements into which composite beings can be broken down, and that form, if not also privation, is an active element, while the substrate is passive.[9] And in keeping with Aristotle's claim that "the principles ought to remain forever,"[10] it also appears that these principles are unaffected by the changes to which they give rise. Thus, the form—or each form, if there are several—fashions the substrate by its presence, or else by its absence allows it to be deformed, while itself remaining eternally unaffected. And even the substrate, though it is said to become and perish, in a sense, by virtue of the presence or absence of this or that form, is nevertheless also treated as a single nature that remains imperishably throughout all these changes.[11]

This initial interpretation of the principles of the beings provides the basis for Aristotle's first response to the perplexity of the early philosophers. For it allows him, as we have seen, to say that things come into being from what is not, in the sense that the substrate is not yet shaped by the eventual form, and also from what is, in the sense that it is at least something even then. This view also allows him to respond, moreover, to a further perplexity of those philosophers. For Aristotle tells us that they went on to say that there is no multiplicity, or that nothing exists except for "that which is itself" (αὐτὸ τὸ ὄν, 191a33), apparently on the grounds that each of many beings would have to be an impossible combination of what is and of some particular determinant, which, not being what is, therefore is not.[12] But if, as our speech suggests, "that which is itself" can be meaningfully understood, in each case, only as that which is precisely some definite being, rather than another,[13] then there can indeed be a multiplicity of beings.

Despite the merits, however, of this account of the principles, its very connection with Aristotle's first response to the perplexity of the ancients points to a difficulty. For the bizarre illustration that troubled us in that response, namely, that of a dog coming into being from a horse, calls our attention to the fact that nothing in this account of the principles would seem to rule out such an event. If the forms are elements whose mere presence in the substrate gives rise to natural beings, there would seem to be no reason why the form of a dog could not supplant immediately that of a horse. For the notion of a single substrate that receives in turn the various forms offers no way of explaining why a certain being must come to be from definite antecedents; or in other words, it offers no way of explaining why the privation that is succeeded by a certain form must be present in the substrate along with some definite form, rather than others. This interpretation of the principles, then, though it may allow us to deny that something can come from nothing, does not rule out, or at least not evidently so, the notion that anything, among the possible beings, can come into being from anything else.[14]

Another difficulty with this account of the principles of the beings is that it leaves it unclear what a being is. To be sure, Aristotle has spoken of the coming to be of each thing (as) what it is called "according to its being" (κατὰ τὴν οὐσίαν, 190b19), thus suggesting that the being of each thing is what it is called, or its form as it comes to light in speech. But his presentation also invites a quite different account of it. For if there is a single substrate that remains imperishably throughout all changes of form, it could well seem, despite our habits of speech, that this is the true being of everything, and that the so-called forms are mere attributes of the one substance.[15] On this view, what come into being would not be beings in the strict sense, but only various attributes of the one persistent being; and Aristotle's account would not differ significantly in this respect from that of the earliest philosophers, at least some of whom allowed for changes in the attributes of the underlying substance.[16] Now clearly enough, Aristotle has been trying to avoid this position. But he acknowledges that he has not simply succeeded in ruling it out by telling us, near the end of his general statement about the principles, that it is not yet manifest whether the form or the substrate is (in the paramount sense) being (οὐσία, 191a19).

Still another difficulty with this interpretation of the principles concerns privation, which Aristotle also calls the opposite to form. Aristotle has said, we recall, that beings do come into being from this third principle, though only by concomitance. His grounds, apparently, for limiting himself to this qualified claim are that the privation or lack of form is only a concomitant in the substrate, and one that ceases to exist there in any sense once the completed being has come to be.[17] But if privation is merely this temporary lack of form in the substrate, one wonders why it is even treated as a principle at all. The mere fact that form cannot come to be present in something without its having not been there before hardly seems a sufficient reason for elevating its initial absence to the status of a principle of becoming. And Aristotle himself suggests at one point that there is no need to speak of a principle opposed to form, since the form itself suffices, by its absence and by its presence, to bring about change in the substrate.[18] Still, he does not follow up on this suggestion, and he continues his account as if it had been established that privation is to be included among the principles. Now this difficulty as to whether, and in what sense, privation is a principle may help to remind us that when Aristotle had first spoken of an opposite to form—before he had even introduced the terms *form* and *privation*—he argued that it was *not* merely by concomitance that something comes into being from its opposite. He also stressed at that time that the true opposite to an ordered arrangement (or "form") is not the mere absence in general of that order, but rather a definite kind of absence, such as the particular manner in which the materials for building a house must first be available.[19] And since a house can be put together only after the materials have been prepared in the appropriate way, it indeed makes sense to speak of this particular kind of absence of its form as a true principle, and not merely one by concomitance, of its coming to be. But by thus helping us to understand the importance of what is opposed to form, Aristotle only adds to the puzzle of why his thematic account of the principles presents a view, or so it appears, according to which this opposite is at most a principle by concomitance.

The difficulties with this initial interpretation of the principles suggest that we should look for another way of understanding them. To this end, it is of help to note that Aristotle has never explicitly even asserted that beings all come to be from a single, persistent substrate (such as might receive, for instance, the form of a horse and then that of a dog). To be sure, he has invited us to assume that he thought so, first by arguing that the passive principle is only one and then by claiming, in the course of the development of his own account, that the substrate, insofar as it is not opposite to the new form, persists throughout the process of coming to be.[20] And yet he has also made it clear that we sometimes speak of a single principle in reference to a number of principles that are one in kind; and though he does later characterize the substrate as one in number, he is referring there, at least primarily, to the particular substrate of an artifact or to a particular being as the substrate of its various attributes. Furthermore, in his first thematic reference to the substrate from which a being, as distinct from a mere attribute, comes to be, he uses as an illustration the seed of a plant or animal, and this clearly is not something that persists.[21] And if the substrate from which a being comes to be need not persist, then we must abandon the interpretation of this principle of becoming as a single something that receives in turn the various forms.

Let me suggest, therefore, that the true substrate from which a being comes to be is in every case something particular and perishable, such as a seed, which in addition to being whatever it first shows itself to be also has the potency to give rise to a new being of some definite kind. This interpretation of the substrate from which a being becomes can help us, as the other could not, to speak properly of unqualified becoming, or the coming into being of a new being, in its distinctness from alteration and the other changes in which a being merely acquires a new attribute (cf. p. 17). And a closer look at what we mean by unqualified becoming confirms the superiority of this interpretation. Later in the *Physics*, Aristotle will tell us that unqualified becoming, as opposed to the other changes, is the coming into being of something that is, i.e., is signified by an affirmative expression, from what is not, or is not signified by any affirmative at all.[22] This account, by the way, makes it all the more difficult for us to be satisfied with his first response to the perplexity of the ancients, according to which coming into being from what is not (understood as privation) occurs only by concomitance (on the grounds that privation is only a concomitant of something that is). But in this same later passage, Aristotle will propose an interpretation of "what is not," an interpretation that builds upon our new view of the substrate, that allows us to see it as a genuine source of becoming. He says there that what is only potentially a being simply or a being at work is in one sense—and in a truer sense than privation, as he also suggests—what we speak of as what is not.[23] Of course, Aristotle does not mean by this claim that a seed, for instance, is not something; but he does mean that what the seed is, above all, is its unfulfilled potency to be the being that has not yet come to be. Accordingly, it makes sense to deny that it is any being in the fullest sense of the word, and even, therefore, to speak of it as what is not.[24] Moreover, as we have noted, it is not merely by concomitance that a being comes to be from something with the appropriate potency. And so my interpretation of coming to be from a particular substrate or potential being has allowed us to understand how unqualified becoming can both require a substrate from which the thing becomes and yet still be a true emergence from what is not.[25] And this interpretation also helps us to see the strength of Aristotle's second response to the perplexity of the ancients, a response that he had told us was based on the distinction between potency and being at work.

Now if the substrate from which a being comes to be does not persist, as at least in the case of natural beings it does not,[26] then the being can not consist of a form in that substrate. Indeed, it does not make sense to describe it as a form in anything else. For what there is, is just the being with various aspects. The form of the being is of course fundamental among these aspects, for it is in terms of form that we give the being its name. And the very fact that we give to it the name of some species—that is, a class of beings whose members are the same in form—shows that there must be other aspects to it as the particular being it is. But none of these other aspects is related to the form as a substrate in which the form exists. And accordingly, we are in a position to begin to resolve Aristotle's question as to whether the form or rather the substrate is being (in the paramount sense: οὐσία; cf. pp. 17–18). For to the extent that we mean by "substrate" something belonging to the being in question, then it now appears that the form *is* the substrate, that is, the being itself, though considered in abstraction from its other aspects. And to overcome this abstraction, or to give a fuller characterization of what the being, or the substrate, is, we may call it a particular instance of that form.[27]

A further advantage of this new interpretation of the principles is that it allows us to understand, as the earlier one did not, what Aristotle can mean in saying that form is an active principle in the production of a natural being. According to the earlier view, we recall, the forms are independent beings that produce embodiments of themselves by somehow becoming present in (a portion of) the substrate, and that eventually cause the perishing of these embodiments by becoming absent from it, while themselves remaining eternally unaffected. On this new view, by contrast, a natural form is the principal aspect of a being that becomes and perishes, and it is this being that has the power to produce others of its kind, as for instance through the production of seeds like the one from which it came itself. And Aristotle helps direct us to this thought by an otherwise puzzling feature of his treatment of the perplexity of the ancients. For he restates the perplexity to include not only the original question of how there can be coming into being from what is or from what is not, but also the question of how what is not or what is can act or be acted upon so as to produce something.[28] And now that we have interpreted form as what (a being primarily) is, we can understand this newest question as a way of asking what it means to say that form acts upon what is not (that being) so as to produce something. Aristotle responds to his question by reminding us that when we say that a doctor, for instance, acts or is acted upon so as to produce something, we mean that he does so insofar as he is a doctor, even though he might also be a builder, a man of fair skin, and many other things as well. By analogy, then, when we say that what is acts upon what is not so as to produce something, we are not thinking of what is, or even of what is something definite, as a form that exists or acts independently. Rather, we mean, in the case of natural beings, that a being, insofar as it is (i.e., is characterized by its form), acts upon what is not, in this sense, but has the appropriate potency, so as to produce another being of the same kind. Thus, a mature animal or plant, insofar as it is characterized by its form, acts upon its nourishment so as to produce a seed of its own kind, and this seed may produce changes in the appropriate material from which a new member of the species comes to be.[29] Or a mass of air, insofar as it is characterized by the form of air, which involves heat, may heat the cooler water beneath it so that it fulfills its potency to be transformed into air.

Now my claim that the substrate from which a being becomes does not persist as a substrate of the being itself, along with the related claim that the form does not, strictly speaking, produce anything, is a modification of what appears on the surface of Aristotle's account. But to say nothing of the hints to which I have already called attention, that surface account is explicitly based on the assumption that there are principles from which natural beings are (constituted as what they are) and that are also those from which they have come into being.[30] And Aristotle presents this assumption in a hypothetical clause, thus helping to call attention to its possible weakness. So we should not be too surprised that the argument as a whole has led us to conclude that there are no such principles. And since the difference between Aristotle's preliminary interpretation of substrate and form and the one that his argument has now led to is largely the difference between their being regarded as imperishable or not, our preference for the latter of these interpretations receives support from a striking suggestion that he makes, in *On the Heaven*, that the principles of the perishable beings may well have to be perishable.[31]

We have now seen, I think, that Aristotle's surface account of the principles (as form, persistent substrate, and privation) and its corollary, the first of his two responses to the perplexity which denied becoming, are not true expressions of his own serious views. But then even apart from the obvious question of why he would say these things at all, there remains the question of why he asserted, after his summary of that surface account, that it was "only in this one way" that the perplexity could be resolved. Did he mean to imply in particular that his second and, as it seemed to me, more adequate response to it, which response does not rely on that account of the three principles, is also not really a resolution (cf. p. 15)? And if he did mean to imply this, what did he find lacking in that response?

Now to try to answer these questions, let us begin by reconsidering the perplexity that surrounds becoming. We recall Aristotle's observation that all the students of nature agree in the opinion that there can be no coming into being from what is not, i.e., from nothing.[32] Yet we know from Hesiod's *Theogony* that it was possible for a thinker of stature to deny this claim. For Hesiod says that at first Chaos, and then Earth, Tartaros, and Eros, came into being; but he does not say, as he does with regard to the subsequent generations, that they came into being from anything or from anyone.[33] But if it is truly an open question whether there can be coming into being from nothing, then for all we know, anything can come into being from anything, and even the assumption that there is nature, or that beings become and perish in accordance with fixed natures, becomes questionable. An indication of what is possible if there is coming into being from nothing is the assumption in the *Theogony*, for instance, that all the gods have come into being but will never perish. In order, then, to avoid having to allow that such things are possible, or in other words to secure the foundations of their study, it would seem necessary for students of nature to give an intelligible account of how coming into being takes place. And it is the apparent impossibility of giving such an account that led those whom Aristotle speaks of as the first, in accord with philosophy, to seek the truth and the nature of the beings, to hold on to their denial of coming into being from nothing by denying that things come into being at all. Now the majority of Aristotle's predecessors in natural philosophy seem to have tried to avoid this paradox by claiming to have identified one or more basic substances (e.g., water, or the atoms), whose nature or natures remain permanently the same, but whose changes in density or whose separations and combinations give rise to the beings that become and perish.[34] Yet Aristotle implicitly rejects all these attempts by saying that his own account of the principles, that is, his own surface account, offers the only resolution of the perplexity. Later, he adds that at least the primary reason for his predecessors' failure to resolve it was their failure to grasp the nature of the (permanent) substrate.[35] For in his view, as it seems to me, the substance or substances on which their doctrines had relied in order to rule out coming into being from nothing do not suffice to make it intelligible, even allowing for separations and combinations and the like, how all the beings of our world, and in particular the living beings, come into being.[36] By contrast, the substrate of his own surface account, since it is characterized above all by a receptivity to and even striving for form,[37] can promise at any rate to play a role in an explanation of the coming into being of these higher beings.

Yet as I have argued, this promise of an explanation is merely that, and there is no permanent substrate in Aristotle's genuine view of the coming into being of things. In his true view, natural beings come from seeds and other such sources, which have the potency to give rise to the various kinds of beings, but which do not persist as substrates of the beings themselves. And this, the second of his responses to the perplexity which denied becoming, has the great advantage, in comparison to the other responses we have considered, of speaking only about the world as we commonly experience it. However, precisely because of its modesty, this response is not even a real attempt to make intelligible how coming into being takes place. It does not even try to say why it is necessary, in normal circumstances, that a given seed should give rise to some definite kind of being. Thus, it does not give grounds for its rejection of the alternative that this "seed" merely ceases to be, and that the emergent being then comes into being from nothing. In other words, since as it now seems, we have only our own limited experience of the world as a guide to tell us what can come from what, we can not be certain that it is impossible for things to come into being from nothing, and the perplexity of the early philosophers remains unresolved.

It is now clear, I think, that the question of why Aristotle did not openly present his true views about the principles of natural beings is bound up with the deeper question of whether those true views include an adequate vindication of the study of nature. And it seems to me, at least, that a chief purpose of his surface account of the principles was to postpone, and even to conceal from many of his own readers, his most serious response to this deeper question. But rather than trying to say more about either of these questions now, I will return to them in a broader context, after first giving more evidence for the legitimacy of my whole approach, in my concluding discussion of Aristotle's manner of writing.

As an appendix to my argument that there are these several levels to Aristotle's thought regarding the principles, I would like to call attention to his discussion of what seems to me to be the somewhat similar case of the philosopher Anaxagoras. To appreciate this discussion, we should first note that it is Aristotle's habit to present his preliminary doctrines as what "we say," that is, what he says as the spokesman for his school. Thus, we have already seen, for instance, that he introduces his first response to the perplexity of the ancients by calling it what "we say" (ἡμεῖς δὲ λέγομεν, 191a34; cf. pp. 13–14). And there are other examples in the *Physics* of this usage. One of the most noteworthy of these examples occurs in his discussion of void, where he introduces the claim that the matter of hot and cold—which in the context means, especially, of air and water—is one in number by saying that this is what "we say on the basis of what has been laid down" (ἡμεῖς δὲ λέγομεν ἐκ τῶν ὑποκειμένων, 217a21). Now in the light of this usage, we are prepared to remark that Aristotle's discussion of Anaxagoras contains an unusual density of references to what "they," that is, he and his followers, say. It is true that he first mentions Anaxagoras in the singular as the author of a doctrine according to which the permanent substrate consists of infinitely many kinds (including all the uniform bodily parts such as flesh and bone, together with the contraries). But when he goes on to speak explicitly of what Anaxagoras thought (οἰηθῆναι, 187a27), as distinct from what he authored (ποιεῖν, 187a24), he uses an ambiguous expression that could mean that he merely "seemed" (ἔοικε δὲ . . . , 187a26–28) to have thought this.[38] And the continuation of Aristotle's account of this view of the substrate, with its further claim that there is some of everything in everything else, presents it only as what "they" say (φασι, 187b1) or as what "they believed" (ἐνόμισαν, 187a36), and there is no reference to Anaxagoras in the singular. Another sign, moreover, that Anaxagoras may not have accepted this doctrine about the substrate is the claim which Aristotle attributes to him, both here and in *On Coming into Being and Perishing*, that coming to be of such and such a sort is alteration.³⁹ For alteration involves the emergence of new characteristics in a substrate, but at least with regard to the characteristics of the infinitely many original kinds, that is precisely what this doctrine is meant to deny. Yet one can understand, on the basis of what we have seen in the case of Aristotle, that Anaxagoras might have taught this doctrine without accepting it. For those who do accept that some of everything, including even flesh and bone and the like, has always been present in every portion of a permanent substrate, are therefore sheltered—at least more than those who believe only in inanimate elements—from doubts regarding their philosophic claim that life could not have come into being from nothing. And yet it seems to me that no genuine philosopher could have accepted this bizarre doctrine.⁴⁰ Thus, I propose that it plays somewhat the same role in Anaxagoras's thought as does Aristotle's own surface account of the principles.

NOTES

1. *Physics* 190b17–29.
2. *Physics* 191a23–b27; 187a26–35.
3. *Physics* 191b20–21; and contrast 190a24–26, 190b4–5.
4. The only support within the ancient tradition for distrusting the surviving manuscripts is a variant reading mentioned briefly by the commentator Simplicius. Simplicius, *In Aristotelis Physicorum Libros Quattuor Priores Commentaria*, in *Commentaria in Aristotelem Graeca* vol. 9, ed. H. Diels (Berlin, 1882), 239.28–30; but contrast 239.18–19. Modern discussions of the passage include *Aristotle's Physics: A Revised Text with Introduction and Commentary*, ed. W. D. Ross (Oxford: Clarendon Press, 1936), 495, and *Aristotle's* Physics, *Books I and II*, trans. W. Charlton (Oxford: Oxford University Press, Clarendon Press, 1970), 80–81.
5. In claiming that this second response is not based on the same three principles, I am presupposing the interpretation of it that I will develop later in this chapter. For one could formulate a response in terms of potency and being at work that would be little more than a restatement of the first one (cf. 217a21–31). On that interpretation, however, it seems odd that Aristotle would speak of it as "another" way of responding to the perplexity.
6. That the second response is the better one is also the view of Thomas Aquinas. See Thomas Aquinas, *In Octo Libros Physicorum*

Aristotelis Commentaria, ed. P. M. Maggiolo, (Rome: Marietti, 1965), bk. 1, lecture 14, par. 126. Aquinas's commentary has been translated by R. Blackwell, R. Spath, and W. E. Thirlkel as *Commentary on Aristotle's* Physics (New Haven: Yale University Press, 1963), 60.

7. *Physics* 190a24–26; cf. *Metaphysics* 1033a5–23.

8. *Physics* 190b17–20.

9. *Physics* 189b16–18, 27–28 (and cf. *Metaphysics* 1014a26–34); 189a20–26, 190b30–35, 191a6–7, and also 192a16–19.

10. *Physics* 189a19–20.

11. *Physics* 192a34–b2; 192a25–34. Aristotle says that the substrate can be considered both as "that in which" (there is privation) and also "according to potency." By this latter expression, he goes on to explain, he has in mind a nature from which something comes into being (and) which is (also) inherent (in the completed thing); and he argues that the substrate in this sense (or matter, as he also calls it) is necessarily imperishable and ungenerated. It is, moreover, this latter view of the substrate that follows most readily from his earlier discussion of it (cf. *Physics* 190a13–25, 192a12–14).

12. Cf. Simplicius, *In Libros Quattuor Priores Commentaria*, 236.1–12.

13. Cf. *Physics* 187a8–9.

14. Compare Plato, *Cratylus* 393b7–c6ff.

15. *Physics* 189a27–34; cf. *Metaphysics* 1029a10–27.

16. *Physics* 187a26–31; cf. *Metaphysics* 983b6–18ff.

17. *Physics* 191b13–17; 190b25–27. cf. Alexander of Aphrodisias, as quoted in Simplicius, *In Libros Quattuor Priores Commentaria*, 238.8–14; Themistius, *In Aristotelis Physica Paraphrasis*, in *Commentaria in Aristotelem Graeca*, vol. 5, pt. 2, ed. H. Schenkl (Berlin, 1900), 30.16–29.

18. *Physics* 191a5–7; cf. 192a14–16.

19. *Physics* 188a31–b21. The definiteness of this opposite to form strengthens the analogy between Aristotle's account of these cases, in which an ordered being first comes to be, and his account of those simpler cases in which change is between contraries in the strict sense (such as hot and cold), and in which privation is accordingly not the mere absence in general of some form, but rather one of the specific contraries in question. Cf. *On Coming into Being and Perishing* 318b14–18, 332a22–23; *Metaphysics* 1055a33–b11ff.

20. *Physics* 189b16–19, 190a13–19. Note, however, that this claim regarding the persistence of the substrate is explicitly based on the premise that we can consider Aristotle's first example of an unmusical man becoming musical as the model for understanding all coming into being.

21. *Physics* 188b36–189a9, 190b23–25; 190b1–5; cf. *On Coming into Being and Perishing* 324b6–7. A further indication that Aristotle does not believe that there must be a single substrate for becoming is his use of the plural αὐτοῖς at 191a1. See also the valuable discussion by W. Charlton, *Aristotle's* Physics, 74–79, 129–45, and also in his "Prime Matter—a Rejoinder," *Phronesis* 28 (1983): 197–211. Charlton fails to recognize, however, how much Aristotle himself contributed to the traditional misinterpretation of his text, and thus he also fails to wonder why Aristotle might have chosen to do so.

22. *Physics* 224b35–225a20.

23. *Physics* 225a20–25, and contrast 225b1–5; cf. *On Coming into Being and Perishing* 317b14–18, *Metaphysics* 1051a34–b1. In this section of the *Physics*, Aristotle speaks of privation as an opposite to form only in cases of qualified becoming, or of motion. In other words, he speaks of change from privation to its contrary form only if it is a change of attributes in a persistent substrate, as distinct from a change from what is not to what is. He even suggests, moreover, that privation can always be signified by an affirmation (as for instance, by "what is naked" instead of "what is not clothed"). See *Physics* 225b1–5, and compare 193b20–21; see also *Metaphysics* 1055a29–b16, esp. a29–30 and b7–8.

24. In the case of the four elements, which come into being from one another rather than from seeds, it is admittedly difficult to treat the coming into being of any one of them as a change from what is not to what is. Yet Aristotle suggests explicitly that it might be correct to do this, at least if the new element is higher, or higher in rank, than the old one (*Physics* 213a1–8; cf. *On the Heaven* 310b11–15; *On Coming into Being and Perishing* 318a35–b33, but contrast 319a29–b5). On the other hand, he does not regard the four elements themselves as beings in the full sense that plants and animals are, and so it is not wholly surprising if the character of unqualified becoming should be less clearly manifest in their case than it is in the case of those other beings (cf. *Metaphysics* 1040b5–10).

25. At *Physics* 225a27–29, Aristotle does not say, as it might appear, that there is coming to be only by concomitance from what is not. What he says, rather, is that even on this supposition (a supposition which he himself has encouraged in book one, and which he might not wish openly to undermine), it is still that which is not that comes to be.

26. An acorn, for instance, does not remain as part of an oak tree, nor does air remain as part of the water that has been formed from it (cf. *On Coming into Being and Perishing* 319b14–18). I am disregarding here the secondary question regarding artifacts, which might be said to

come into being from the reshaping of a substrate—such as wood, for instance, or bronze—which persists at least as long as the new beings do. Even in these cases, however, I do not think that the being is appropriately characterized as a form in a substrate. cf. *Metaphysics* 1045a7–b23 and see p. 17.

27. Consider *Physics* 188b16–21, and cf. pp. 18 and 85–86. See also Charlton, *Aristotle's* Physics, 71–73.

28. *Physics* 191a34–36.

29. See *On the Generation of Animals* 724a14–726a28, 729a34–730b32, and throughout.

30. *Physics* 190b17–23; cf. pp. 15–16.

31. *On the Heaven* 306a9–11; consider, again, *Physics* 192a25–b4.

32. *Physics* 187a32–35. The reference, at 191b35–192a1, to a Platonic claim that there is coming into being from what is not should be compared, rather, to Aristotle's own later suggestion that what is not may be understood as what is only potentially a being in the full sense. See, again, 225a20–25.

33. Hesiod, *Theogony* 116–22.

34. *Physics* 187a12–26; *Metaphysics* 983b6–984a29.

35. *Physics* 191b30–36.

36. Cf. *On Coming into Being and Perishing* 333b3–20, 335b24–336a12.

37. *Physics* 192a16–19.

38. To be sure, the simpler interpretation of this sentence is that it "seems" to be for the two reasons that Aristotle here proposes that Anaxagoras thought that the substrate was thus infinite (without there being any further suggestion that it only "seems" to be the case that he did think so). But the reading that I have suggested is also possible, and it seems preferable in the light of the other factors discussed in the text.

39. See *Physics* 187a29–30. That this is a reference to Anaxagoras (and his followers) is confirmed by the explicit statement in *On Coming into Being and Perishing*, 314a13–16, that Anaxagoras did identify coming into being and perishing with alteration. It is true that there is no evident support for this statement in the fragments that remain to us from Anaxagoras. Yet I see no reason to distrust what Aristotle says on this matter. And his use in both passages of the verb καθέστηκε(ν) strongly suggests that he had a specific Anaxagorean text in mind. It is worth noting that the passage in *On Coming into Being and Perishing* stresses the inconsistency between Anaxagoras's claim that becoming is alteration and his doctrine about the substrate.

40. Consider the difference between the singular and plural subjects in *On Coming into Being and Perishing* 314a24–b1. See also Alexander of Aphrodisias, *In Aristotelis Metaphysica Commentaria*, in *Commentaria in Aristotelem Graeca* vol. 1, ed. M. Hayduck (Berlin, 1891), 68.5–70.9.

Newton's claim, however, about the origin of the solar system appears only in a General Scholium that he appended to his *Principia* twenty-six years after its first publication. And these theological reflections play no role in the body of the work, just as they have played no role in the subsequent history of modern physics.

More generally, modern science from the beginning has sidestepped the question of ends or purposes in nature, while encouraging people to take less and less seriously the possibility of their existence, by trying to formulate laws that would encompass as many phenomena as possible without any reference to such ends. But it has not succeeded, of course, in encompassing all phenomena, at least not yet. And even in the case of those phenomena that it does encompass, the claim that it has understood them adequately is dubious. For its laws are mathematical idealizations, idealizations, moreover, with no immediate basis in experience and with no evident connection to the ultimate causes of the natural world. For instance, Newton's first law of motion (the law of inertia) requires us first to imagine a body that is always at rest or else moving aimlessly in a straight line at a constant speed, even though we never see such a body, and even though according to his own theory of universal gravitation it is impossible that there can be one. This fundamental law, then, which begins with a claim about what would happen in a situation that never exists, carries no conviction except insofar as it helps to predict observable events. Thus, despite the amazing success of Newton's laws in predicting the observed positions of the planets and other bodies, Einstein and Infeld are correct to say, in *The Evolution of Physics*, that "we can well imagine that another system, based on different assumptions, might work just as well." Einstein and Infeld go on to assert that "physical concepts are free creations of the human mind, and are not, however it may seem, uniquely determined by the external world." To illustrate what they mean by this assertion, they compare the modern scientist to a man trying to understand the mechanism of a closed watch. If he is ingenious, they acknowledge, this man "may form some picture of a mechanism which could be responsible for all the things he observes." But they add that he "may never be quite sure his picture is the only one which could explain his observations. He will never be able to compare his picture with the real mechanism and he cannot even imagine the possibility or the meaning of such a comparison."[4] In other words, modern science cannot claim, and it will never be able to claim, that it has the definitive understanding of any natural phenomenon. And accordingly, we should not allow ourselves to be so dazzled by it as to suppose that it has refuted Aristotle's teaching about natural ends.

Let me turn, therefore, to an examination of that teaching itself, to try to see whether in fact it makes sense. I begin with a brief summary of Aristotle's doctrine of the four causes, among which the end or final cause is included. In trying to explain anything, or to answer the question Why? about it, there are four ways, according to Aristotle, that we can respond. These four kinds of responses, these four ways of saying "because," are each a kind of cause of the matter in question. In one sense, we speak of the cause of something as the constituent or constituents from which it comes into being and of which it is composed. Thus, for instance, we call the bronze in a statue a cause of it, because the statue could not exist without the bronze that it is made of.[5] In another and more important sense, however, we say that the cause of something is its form, since it is the form, rather than the constituent material, that explains why the being is what it is. Thus the bronze constitutes a statue because it has the form of a statue, not because it is bronze. And by the "form" of a being we do not mean its visible shape alone, but rather its whole character—a character that we may not be able to articulate clearly at first, but that we must already understand to a considerable degree simply to identify the being as what it is. To take our example again, we call a statue a statue because of its form, by which we do not mean its shape alone but, more importantly, its being the sculpted image of some other being, such as a man or a god. Similarly, we call a human being a human being because of its form, by which we mean above all that we are animals with the power of reason. Now in addition to the constituent cause and the formal cause, a third kind of cause, which helps to explain coming into being as well as all other changes or motions, is the source from which a given change begins. Hence, for example, the sculptor is a cause of the statue's coming into being, and therefore also, in a sense, of the statue itself, as the parent is a cause of the birth of its offspring and the wind is a cause of rustling in the leaves. This principle of change, which Aristotle sometimes calls the moving cause, can also, however, be a cause of a state of rest, as when a man decides to stop running and to remain still. Finally, a fourth kind of cause is the end or purpose for which something comes into being or for which it exists. Thus health, for instance, can be a cause of walking by being the purpose for the sake of which someone might take walks. Or to return to our first example, giving honor to a god or perhaps pleasure to the human beholders might be the end—or the final cause, as it is also called—for the sake of which the sculptor produced the statue.

To focus specifically now on final causality, the examples I have just given suggest a definite manner in which the end of a motion operates as a cause. First, it is present in thought to the agent who begins the motion, and it is thought of as being good, and as being attainable by his own efforts. And as a consequence, the agent proceeds to try to realize in deed the end that he originally had in mind. In these cases, then, the final cause is itself a factor, together with what we have called the moving cause, in initiating the motion that leads up to it as an end. And most people would agree, I think, that this is a true account of such motions as that of a statue being sculpted, for it is the craftsman's anticipation of the completed work that directs his hands and tools. But Aristotle says explicitly that the end is a cause of natural motions, such as the growth and reproduction of living beings, and not only of motions initiated by human art.[6] Indeed, he justifies his use of analogies from the arts in the explanation of natural processes by claiming that art imitates nature; and he means by this not merely that some arts are representational but, more broadly, that all the arts derive their own directedness toward ends from that already to be found in nature.[7] He even goes so far as to speak of Nature herself as a kind of artisan who makes all things for a purpose and nothing in vain.[8] Yet we have no clear evidence to suggest that nature is an intelligent being; and since to act for a purpose would seem to presuppose intelligent forethought, it is hard to understand how nature could act in this manner, or how natural motions could be for the sake of any ends.[9] So couldn't it be, instead, the result of accident that natural processes often turn out in a manner that seems to suggest motion for the sake of an end? Now this is not, as it might appear, a question that only we in the post–Darwinian age can ask, but in fact a central question that Aristotle explicitly confronts throughout his account of purposes in nature. To see how he responds to it, let us look more closely at the text of the *Physics* itself.

Aristotle discusses accidental causation immediately after his account of the four causes, and he indicates the tremendous scope of the question about it by noting that some men had attributed the existence of this world, and indeed of all ordered worlds, to chance. Now by "this world" Aristotle means, not the hypothetically expanding universe of contemporary science, but rather this perceptible world, the world in which we live, along with other animals and plants, upon the earth and under the heaven. The beauty and grandeur of this world have always aroused in men a sense of wonder, and many have felt, as Newton did, that it must be the product of intelligent forethought.[10] Yet according to some of the early philosophers, this world of ours came into being, as did many other worlds, merely by chance. Now in response to these philosophers, Aristotle begins by saying that their claim is worthy of wonder. For those, he says, who attribute the world and the most divine of the manifest beings (i.e., the stars) to chance do not speak of animals and plants as coming into being by mere luck, but rather by nature or mind or some other such cause, on the grounds that each kind of seed typically produces some definite kind of result. And on precisely these grounds, he continues, what they say about the world as a whole is especially strange, since we see nothing in the heaven coming into being by chance, whereas we see many things resulting from luck among the perishable beings here on earth.[11] Now at this point, it would certainly be possible for us moderns to raise objections to Aristotle's argument. But I would suggest that we defer our objections and consider instead that Aristotle has not said, at least not yet, that those who attribute the world to chance are wrong, but merely that their assertion is wonderful and strange—as I venture to say it is. And he adds that if what they say is true, this very fact deserves careful attention, as indeed it does. Rather than assume, then, that Aristotle has already rejected the view in question about the origin of the world—whether on adequate or on inadequate grounds—we would do better to follow him as he goes further into the matter by investigating more generally what chance is and what its relation is to nature.

Aristotle begins his account of chance by first discussing what he calls luck and by observing that we speak of luck, not as a cause of those things that come into being always or for the most part in the same way, but rather as a cause of those that come about contrary to any such necessary or usual patterns. Yet it is not simply the unusualness of an occurrence that makes it a case of luck. It is also necessary that it be the kind of occurrence that *might* have been brought about for the sake of something and from intelligent choice. Such a result, when it comes about, however, merely by concomitance, is what is properly said to be from luck. Aristotle offers an example to help clarify what he means in saying that luck is a cause of its results by mere concomitance. A certain man, he says, would have gone to the marketplace to recover the money he was owed if only he had known that his debtor was there. But he did not know, and so he did not go to the market for that purpose; it was merely a concomitant of his going for some other reason that his trip served the purpose of recovering his debt. Thus, the result came about from luck, and the man is said to have gone to the marketplace from luck. If, however, he had chosen to go to the market for this purpose, or if he went there always or for the most part when collecting on his debts, neither his going nor its result would be from luck. To speak of luck as a cause, then, is not to deny that there must always be some definite, particular cause of a lucky result. Indeed, in keeping with the fact that luck is responsible only for the kinds of outcomes that might have resulted from intelligent choice, only rational agents, who alone are capable of such choice, can be said to act from luck. But we attribute the result to luck when the agent's action merely happens to serve a purpose without his having intended it to do so.

After preparing the way with this account of luck, which has the merit of being immediately intelligible, Aristotle turns to the discussion of chance, or spontaneity, as we might also translate the Greek term. Chance is a broader class that includes luck as a subclass. Like luck in particular, chance in general belongs to the overall category of moving causes.[12] And it too is a cause of something that, while it is the kind of thing that might have been brought about for a purpose, does not in fact occur for the sake of what results. The greater extension of the term *chance* is due largely to the fact that the particular cause of what we call a chance outcome need not be an agent capable of intelligent choice, as it must in the case of luck, but can be any being, whether animate or even inanimate. Yet Aristotle suggests more of a difference between chance in the wider sense and luck in particular than this. He also suggests that chance results need not be of such a kind as even *normally* to require an intelligent agent, but that in some cases they are ones that might have been brought about by nature, where nature is understood as a power, distinct from intelligence, that nevertheless also acts for a purpose.[13] This suggestion, however, is a difficult one. For as I have already mentioned, it is hard to see what it can mean to act for a purpose if not to act from intelligent forethought, and so it is hard to understand what that nature could be that Aristotle is contrasting with chance. The examples, moreover, that he gives as illustrations of chance serve to highlight this difficulty. Let me mention only one of these examples, in which the difficulty becomes explicit. Aristotle says that a stone fell by chance when it hit a man who was passing by below; and the reason he gives for attributing its fall to chance is that it might have fallen through someone's agency, and for the sake of striking the man, although in fact it did not. In other words, to explain what he means by a chance event, he contrasts it, not with one that might have come about from nature, but with one that might have been chosen by an intelligent, if in this case an unfriendly, being.[14]

Now Aristotle is aware that his examples do not fulfill our expectation that chance results are sometimes to be contrasted with what nature, as distinct from intelligence, might have brought forth. And so he adds that chance in the wider sense is most distinct from luck in the sphere of natural generation, when an animal or plant comes into being contrary to nature, that is, maimed or with some deformity. In such cases, he says, we do not speak of luck, but rather of chance—and this not merely, it seems, because these beings are not produced by an agency capable of intelligent choice, but also because the nature that might have produced them, and done so for a purpose, is itself not an intelligence. Aristotle adds, on the basis of a consideration that I will return to later, that these impairments and deformities are not precisely even from chance. Nevertheless, his reference to them serves to support the suggestion that chance, as distinct from luck, can be a cause of what might have arisen from the purposive action of nature.

The question remains, however, of what it can mean for nature to act for a purpose. And Aristotle again calls our attention to this difficulty, in the course of a renewed attempt to answer the question about the cause of the world. What he says is this: since chance and luck are causes by mere concomitance of what either mind or nature might have caused, they cannot be prior to mind and nature themselves; so that even if chance is responsible for the world, mind and nature are necessarily prior causes, both of many other things and of this whole.[15] Now it is not immediately clear, of course, how Aristotle arrives at this conclusion or how it is to be understood. But what I would first like to emphasize is his claim that *both* mind *and* nature must be prior to chance, even on the assumption that chance is responsible for the world. That he makes this claim about mind, in particular, comes as a surprise. For the assumption about a chance origin of the world can refer, and almost certainly does refer, to chance in the wider sense, as distinct from luck. After all, the philosophic position to which Aristotle is responding claims that our world came into being through the blind interaction of the elements, not from the lucky or unlucky actions of an intelligent being. Yet if we assume that chance in this sense is responsible for the world, one might have supposed, perhaps, that nature is a still prior cause, on the grounds that chance is a cause of what might have been produced by nature, which must, therefore, already be at work. But what Aristotle says, to repeat, is that nature *and* mind must *both* be prior causes of the world. And if mind, as well as nature, must already exist for there to be chance in the wider sense, this would seem to imply that nature is not capable of acting for the sake of ends—those ends that are presupposed whenever we speak of chance—unless it is somehow accompanied by mind. And however much this view of nature might help us to understand how nature can act for a purpose, it is at odds with Aristotle's earlier suggestion that it can do so on its own.

And yet perhaps this new suggestion, that nature depends on mind in order to do its work, is an intentional modification of the earlier one; and in fact, the very next chapter tends to support this modified view of nature. Aristotle says there that a natural being has two sorts of moving causes, one of which is not natural itself, since it is wholly unmoved. He then identifies this unmoved mover with the being's end, or final cause, and also with its form; but the form he is referring to cannot be the form as it exists in perishable beings, since we have already been told that these forms are perishable themselves, and hence not entirely unmoved.[16] The reference must therefore be to some other kind of form, and the only other kind that Aristotle believes to exist is the form as it exists in thought, whether in human thought or divine thought. If natural change, then, is to be initiated in part by final causes that are also unchanging forms, it seems that these forms must be present in a divine mind, a mind that somehow solicits natural beings to move toward the perfection of their exemplars.[17] And indeed, Aristotle has already suggested something like this earlier in the *Physics*, where he said that the two highest principles of nature include one that is divine and good and aimed for and another that naturally aims for and reaches toward the first one.[18] And there are quite a few passages scattered throughout Aristotle's works that speak of a divine mind as the Prime Mover and ultimate source of all life and nature.[19]

And yet despite the attractiveness of this suggestion about the subordination of nature to a divine mind, it does not yet allow us to see how natural motion as such can be for a purpose. For purposive motion, at least as we understand it, presupposes more than the mere existence in some mind of the form that is to be aimed for as its end. It also presupposes that this mind be that of an agent which can set itself in motion with a view to that end. Thus, for instance, if the heavenly spheres are intelligent, living beings, as Aristotle says they are, it is at least conceivable that they might set themselves in motion in order to approximate the perfection of the divine forms.[20] But unless even the seeds of natural beings here on earth are self-moving, intelligent agents, or unless they are tools in the hands of some other such agent, the assumption that the forms exist in a divine mind does not yet make it intelligible how natural growth, for instance, can be purposive. Perhaps, then, the claim that led us to these speculations about the divine mind, the claim that *both* mind *and* nature must necessarily be prior to chance, allows for another interpretation. In particular, in saying that *mind* must be prior to chance, even if chance is responsible for our world, Aristotle may be trying, not to help us to see how *nature* could act for a purpose, that is, in conjunction with mind, but rather to indicate something about mind and chance in themselves.

And in fact, there is such a way of interpreting the claim about mind being prior to chance. For according to Aristotle, there are different senses of the term *priority*. In particular, if knowledge of one thing is presupposed by that of another, the former is said to be prior in terms of knowledge, without necessarily being prior in other senses, such as that of existing earlier in time.[21] Thus, if we assume, as Aristotle invites us to do, that the world emerged through chance, it might still be true, as he claims, that mind must be a prior cause of it; but perhaps this is only because of priority in terms of knowledge. At all events, Aristotle would have good grounds for saying that mind is prior to chance in this sense, for chance results are defined as those that might have been brought about for a purpose, and we understand purpose as that for the sake of which an intelligent being would act. Knowledge of chance, then, presupposes a prior knowledge of mind, or, more precisely, of ensouled minds, since it is these that are capable of action for the sake of ends.[22] Moreover, if being in the paramount sense is what is true, or what is truly understood by some mind—and Aristotle makes this claim in the *Metaphysics*—then this priority of mind to chance in terms of knowledge also implies a certain priority in terms of being.[23] Chance would not be fully what it is if it were not understood—understood, that is, in its relation to the ends pursued by intelligent beings. And this is true even of the chance, on the hypothesis that it is chance, that was first responsible for the coming into being of our world.

In trying to account for the priority of mind to chance, I said that the definition of chance contains the notion of purpose and that purpose is understood as that for the sake of which an intelligent being would act. And yet we cannot forget Aristotle's earlier suggestion that what is for a purpose is what might have been done *either* from intelligence *or else* from nature.[24] Thus, it seems that I have been overhasty in claiming that we need to understand intelligence, as distinct from nature, in order to understand chance. However, the difficulty of even conceiving what it can mean for nature to act for a purpose has been our chief puzzle all along. And so let me suggest that natural ends themselves, like the ends that just happen to result from chance, can not be understood without a prior understanding of those ends that are pursued by intelligent agents. Now it is true that when Aristotle first speaks of purpose in the *Physics*, he makes no explicit reference to intelligence. What he says, rather, is that "nature," by which he means the form of a natural being, "is an end and [that] for the sake of which; for in the case of those beings whose motion is continuous and has some end, this last stage is also that for the sake of which."[25] And yet he has to add immediately that not just *any* last stage of a being's continuous motion is a true end or purpose; after all, he observes, it was ridiculous of the poet—presumably a comic poet—to say that death was the purpose for which a certain man had been born. Aristotle goes on to explain that only the best kind of last stage wishes to be an end; and he speaks repeatedly of ends or purposes as being, or as wishing to be, what is better or best.[26] But to be better or best presupposes a range of alternatives as well as some being with the intelligence to compare these alternatives and to judge among them. Accordingly, all knowledge of what it is to be an end, including a natural end, depends on knowledge of the features of intelligent choice.

My argument has now shown, I think, how Aristotle could justify his claim that even on the hypothesis that our world emerged through chance, mind would still be prior in causality to chance. But it has not made sense of the other half of Aristotle's claim, namely, that nature would also, on this hypothesis, be prior to chance. For I have not relied on the assumption that nature, as distinct from mind, acts for the sake of ends. But unless nature does act in this way, it is not even clear how it differs from chance, let alone how it must be prior to it. After all, chance happenings themselves are of such a kind that they *might* have come about for the sake of ends. Now there is at least one difference between nature and chance, which I alluded to earlier, when I said that congenital impairments and deformities in natural beings are not precisely from chance. For Aristotle's definition of chance included a criterion that I have not yet mentioned, namely, that the causes of chance results must be external.[27] Congenital impairments and deformities, by contrast, are said to have an internal cause. And it is not only such deformities in natural beings that can be distinguished from chance results on the basis of this criterion. Their normal development is also said to originate from a cause within themselves.[28] Thus, the normal growth of living beings is said to be from nature rather than from chance, since it begins from a parental seed, which has within itself, as the parent did already, the power to initiate those motions that lead to a mature animal or plant. On the other hand, Aristotle thinks, rightly or wrongly, that in some species of living beings the usual mode of generation is spontaneous, i.e., from chance. Such generation comes about, he says, because the material, such as putrefying slime, from which these animals or plants begin their development can be moved by itself with the kind of motion that more typically must be caused in maternal residue by a paternal seed.[29] In generation from chance, then, the original source of motion is something other than the kind of being that results from it; and in general, we can distinguish nature, as an internal source of change, from all external causes such as chance. Still, however, this criterion by itself does not suffice to explain why natural motion is said to be for a purpose. And neither does it allow us to understand Aristotle's claim that nature must be prior to chance. Indeed, the view that nature is only an internal cause of motion adds an additional reason for doubt on this score. After all, the survival, growth, and reproduction of natural beings are dependent on a great many external conditions. Accordingly, nature, or at least the nature of each kind of perishable being that we know of here on earth, must depend for its very existence on something else—either on some higher principle of order or else on chance.[30] And on the hypothesis that our world as a whole emerged through chance, chance would seem to be prior to nature, rather than the other way around.

If, moreover, one holds that the world as a whole emerged through chance, it becomes all the harder to see how there could be a purpose behind the internal development of particular natural beings. And some of Aristotle's philosophic predecessors had in fact argued that there was none. Aristotle recasts their argument in the form of a perplexity, as follows. Zeus does not cause rain, the argument begins, in order to make the crops grow, any more than it rains in order to damage the crops on some poor farmer's threshing-floor; rather, the rain falls by necessity as the evaporated moisture rises and then cools, and it just happens to be advantageous or harmful for human beings. And if such blind causality is the truth behind the apparently benevolent order of the world as a whole, what prevents there being the same kind of causality at work in the formation of particular natural beings? This argument acknowledges that the parts of organic beings are suitably arranged for the activities through which they survive and reproduce. But it claims that these arrangements first originated from blind necessity, as did many others, and that it merely turned out that these were serviceable, as the others were not, for survival and reproduction. The animals and plants that we see around us continue to survive because their parts turned out *as if* they had been organized for a purpose; but they were not organized for any purpose in fact.

Aristotle admits that one might be perplexed by this attempt to explain the apparent purposiveness in the structure of organic beings, but he denies that things could have happened in the way that it suggests. He offers in rebuttal a whole series of arguments that nature is truly a cause for the sake of something. And the first of these arguments at least suggests that natural purposiveness is not limited to the internal structure of organic beings. For instead of granting the premise that the causes of rain are indifferent to good and bad, and arguing that the parts of living beings are nevertheless arranged for a purpose, Aristotle lays down premises which imply that even the rain—or at least its seasonal patterns—is for the sake of something.[31] To this extent, then, he proceeds as if he assumed that it does not make sense to think of nature as acting purposefully even within a limited realm unless the world as a whole is ordered for the sake of ends.

Unfortunately, however, this first in Aristotle's series of arguments for natural purpose is extremely problematic. And yet partly for this very reason, it is of such importance that I wish to go through it in detail. Aristotle begins by saying that "these things, and indeed all things that are by nature, come into being either always or for the most part in the same way," but that "nothing that is from luck or from chance does so." In support of this assertion about luck and chance, he adds that "it is not believed to be from luck or from coincidence"—which latter term appears to be a synonym for chance—"when there are frequent rains in winter, but [only] if they come during the dog days [i.e., in summer]." Similarly, he says, "heat during the dog days is not [believed to be from luck or from coincidence], but [only] if [it happens] in winter."[32] Now it is noteworthy that the explicit basis for Aristotle's claim that normal weather patterns are not from coincidence is what is commonly believed. For those who make the argument against him, and who speak of a chance origin of the various species of living beings, would object that common opinion has too limited a time frame, and that the patterns whose regularity it is struck by exist only for a short time, in the light of eternity, as does our world itself.[33] They would say that even if the persistence of these regular patterns is not merely coincidental, but a necessary consequence of the elements in their juxtaposition, the whole order within which these patterns exist is the temporary and unintended aftereffect of an original coincidence. And yet despite the obvious fact that this is a possible objection, Aristotle does not even attempt here to justify his reliance on the common perspective regarding the permanence of our world; and by being explicit about this reliance, he calls attention to the question of whether that perspective is adequate.

Aristotle continues his argument as follows: "If, then, [things] are believed [to be] either from coincidence or for the sake of something, and if it is not possible for these things to be either from coincidence or from chance, they would be for the sake of something."[34] Now this conclusion follows, of course, if its premises are sound. But the premise that things must be either from coincidence or else for the sake of something is open to doubt. For what about the bad things that happen always or for the most part? In particular, what about the death that comes inevitably to all living beings here on earth? The fact that birth is followed by death is no mere coincidence, if anything is not, and yet it is hard to see a purpose for it. Aristotle helps us, moreover, to see the dubiousness of the claim that being from coincidence or else being for a purpose is an exhaustive alternative by explicitly basing this premise too on common opinion. And a further sign that he is aware of its weakness is that he does not go on to claim that *everything* by nature is for a purpose, even though he had begun his argument by asserting that everything by nature comes about always or for the most part in the same way. For if nothing that happens with such regularity can be from coincidence, as he asserts, and if being from coincidence and being for a purpose are indeed the only alternatives, it would follow that everything by nature—including rainfall in winter, as I mentioned earlier—*is* for a purpose. And yet, to repeat, Aristotle does not even claim that this is true.[35] What he says, rather, is that if the particular things in question, that is, the arrangement of parts in living beings, cannot be from coincidence or from chance, then *these* things would be for a purpose. By thus retreating from the full implications of his argument, Aristotle suggests that this more limited conclusion, regarding the parts of organic beings, is not meant to rely wholly on the reasons that have been made explicit. Finally, Aristotle concludes his overall argument by observing that all such things are by nature, so that there is purpose, or "the for the sake of something," among the beings that come into being and that are by nature. But this final conclusion, since it is based on the previous one, can also not be justified by the explicit argument alone.

The disregard, in Aristotle's explicit argument, of the problem of natural evils is closely related to its other difficulty, its failure to respond to his adversaries' contention about the origin of the world. For whatever weaknesses there may be in the doctrine that our world order, and in particular its species of living beings, emerged through chance and necessity alone, this view is the one most obviously compatible with a clear account of death and other natural evils. And so by pointing to his awareness that he has failed to acknowledge these evils in his own argument, Aristotle calls still further attention to his having left it open, at least for now, that his adversaries might be right about the question of ultimate origins. Moreover, in a subsequent argument against the claim that natural processes take place simply from necessity, he says that the earlier stages of such development do not lead necessarily to its culmination, but that the necessity in these cases is merely hypothetical, in the sense that there must be the appropriate preconditions if a being with a given nature is going to exist.

He suggests, in other words, that there is no unqualified necessity, either from material causes or otherwise, for the existence of any natural being, and thus he again calls attention to the possibility that the world as a whole might owe its existence to chance.[36]

But what, then, is the basis for Aristotle's emphatic assertion, throughout this whole series of arguments and elsewhere, that the end or purpose is a cause of natural development? For if there is no evidence that such development results from the forethought of a self-moving agent, and if the possibility must even remain open that life itself first emerged through chance, how can he speak with such confidence of purposes or ends in nature? Now to answer this question, we must begin by noting that whatever the origin of life, if it had an origin, and whatever the more immediate moving causes of natural development may be, it remains true that this motion issues regularly, throughout human experience, in living beings whose parts work wonderfully well together in carrying out their life activities. These activities, such as nutrition, reproduction, perception, locomotion, and thought, are the manifestations of the form, or soul, of the being in question; and this form cannot be understood in the light of its material conditions alone. Rather, it is only in the light of the form that the material conditions can be seen as what they are. The bodily organs of a living being must be seen as instruments for performing its characteristic activities, activities without some of which, at least, the body itself would cease to exist; and even the more ultimate conditions of life must be understood, at least in part, in the light of their contribution to this end.[37] Accordingly, even if the term of natural development is not in general anticipated by its moving cause, as it would have to be if it were a purpose in the strict sense, the notion of purpose, or of what something is good for, remains central to the proper understanding of natural beings. And since living beings become what they most truly are by attaining their mature form, their typical growth toward maturity must also be seen in the light of this end. This is not to deny that mature development may indeed result from the same kind of blind necessity, given certain conditions, as does deformity or premature death, given certain others. But as words like *deformity*, *mature*, and *premature* themselves already suggest, we nonetheless see these outcomes either as successes or as failures of a tendency toward a natural goal. We can disregard, or pretend not to notice, this privileged status of the mature form; but it remains an inescapable fact of the world as it presents itself to our experience.

Now these conclusions, to repeat, are independent of the question whether the world as a whole emerged through chance. And thus the centrality of form to the understanding of natural beings also allows us to make sense of the claim that even if the world did emerge through chance, nature would still be a prior cause of it. For the world consists primarily of natural beings, and of living beings in particular, which are properly seen as what they are in the light of their forms. Hence knowledge of these natural forms, and of these forms as the normal ends of growth, is presupposed by the very question of whether the beings they characterize, or even the world as a whole, first originated through chance. Nature as form and as the end of motion is necessarily prior to chance, at least in the sense that chance can not be understood without a prior knowledge of this nature, just as we saw earlier that it could not be understood without a prior knowledge of the features of intelligent choice.[38]

This notion of the mature form as the end of natural development is a far cry, of course, from the strict sense of purpose, according to which the end is anticipated by thought and thus helps to initiate the motions that lead up to its realization. And although my account of natural ends has the advantage of being compatible—as is the bulk of the *Physics*—with the possibility that our world emerged through chance, the fact remains that the final book of the *Physics* and the first book of *On the Heaven* contain elaborate arguments for its being eternal. Since Aristotle also claims, moreover, that the highest principle of the world is a divine mind, he still seems to hold out the hope that natural motion is somehow also activity for a purpose in the strict sense of the word. Yet as I suggested earlier, I do not think that Aristotle himself could have given credence to the notion that anything other than a self-moving, intelligent agent—which the divine mind of his theology is not—can act for a purpose.[39] Accordingly, I contend, his apparent openness to the view that natural motion in general has purposes in the strict sense is a deliberate accommodation to popular views.

NOTES

1. Jacques Monod, *Chance and Necessity*, trans. Austin Wainhouse (New York: Vintage Books, 1972), 21. The claim made in this passage is significantly qualified on p. 41 (which should be read in connection with p. 104, along with the slightly more tentative remarks on pp. 112–13). The original claim is restored, however, on pp. 176–77. Note also the quotation marks around the word "true" on pp. 21 and 176.

2. *The Philosophical Works of Descartes*, trans. Elizabeth Haldane and G. R. T. Ross, vol. 1 (London: Cambridge University Press, 1972), 173; contrast, however, p. 194.

3. *Principia,* vol. 2, *The System of the World*, Motte's translation revised by Florian Cajori (Berkeley and Los Angeles: University of California Press, 1934), 543–44. Compare p. 546 and pp. 668–70.

4. Albert Einstein and Leopold Infeld, *The Evolution of Physics* (New York: Touchstone Books, 1966), 31.

5. The fact that in the case of *natural* beings the cause from which they come into being does not persist as a constituent of the beings themselves is irrelevant to this discussion of final causality. Cf. pp. 20, 22.

6. See, for instance, *On the Soul* 415a22–b21; *On the Parts of Animals* 641b23–33; *On the Generation of Animals* 730b8–32; *Physics* 199a20–b14.

7. *Physics* 194a21–22, 199a15–17.

8. *On the Parts of Animals* 641b10–12; *On the Heaven* 291b13–14, 271a33; *Politics* 1256b20–22; *Posterior Analytics* 94b36.

9. I will sometimes use the term *intelligent* in a wide enough sense so as not to exclude the intelligence of the irrational animals, since they all appear to be capable of self-motion for the sake of ends which they sense, imagine, or even somehow understand (*Physics* 199a20–23; cf. *On the Motion of Animals* 700b15–701b1). However, Aristotle clearly limits the capacity for choice to rational animals, and so I will speak of intelligent *choice* only in reference to them.

10. See *On the Parts of Animals* 641b10–23; compare Plato, *Philebus* 30a9–c8; *Laws* 885e7–886a4.

11. *Physics* 196a24–b5.

12. *Physics* 198a2–4.

13. *Physics* 196b21–22, 198a3–6; cf. 197a5–6 and 197b18–22.

14. Cf. *Aristotle's* Physics, *Books I and II*, trans. W. Charlton (Oxford: Oxford University Press, Clarendon Press, 1970), 109–10.

15. *Physics* 198a5–12.

16. *Physics* 192a34–b1; cf. 198b1–4 (but contrast 224b5–6).

17. *Physics* 194b26; Compare Charles H. Kahn, "The Place of the Prime Mover in Aristotle's Teleology," in *Aristotle on Nature and Living Things: Philosophical and Historical Studies Presented to David M. Balme*, ed. A. Gotthelf (Pittsburgh: Mathesis Publications, 1985), 182–205. See also Thomas Aquinas, *Summa Theologica* 1, Q.2, A.3; 1–2, Q.40, A.3.

18. *Physics* 192a16–19.

19. See, for instance, *Metaphysics* 1072b13–14; *On the Heaven* 279a22–30.

20. *On the Heaven* 292a14–22, 285a29–30; *On the Motion of Animals* 700b25–32; *Metaphysics* 1072a19–b4, 1073a22–b1.

21. Compare *Metaphysics* 1018b14–37; *Categories* 14a26–b8.

22. Compare Wolfgang Wieland, *Die aristotelische Physik: Untersuchungen über die Grundlegung der Naturwissenschaft und die sprachlichen Bedingungen der Prinzipienforschung bei Aristoteles*, 2nd ed. (Göttingen: Vandenhoeck and Ruprecht, 1970), 259–60.

23. Contrast *Metaphysics* 1051a34–b6 with 1027b18–1028a6. See also *On the Soul* 426a15–26, 430a19–22, 431a1–4; and compare *Physics* 223a16–29. In the light of this need for mind if there is to be being in the fullest sense, one can also interpret the claim that mind would be a prior cause of this whole, even on the hypothesis that it first came to be from chance, as a claim that what is responsible for the completeness of this world is a more important cause than that of its mere beginnings.

24. *Physics* 196b21–22; cf. n. 13.

25. *Physics* 194a28–30ff.

26. *Physics* 194a32–33, 195a23–26, 198b8–9, 198b16–17.

27. *Physics* 197b18–22, 197b35–37; cf. 196b34–197a2.

28. *Physics* 192b8–32, 199b14–17; and contrast 197a1–2.

29. *Metaphysics* 1034b4–7; cf. 1032a25–32; *On the Generation of Animals* 715b25–30, 762a8–763b16; *History of Animals* 539a21–25, 547b18–23.

30. Consider *Posterior Analytics* 95a3–6.

31. Cf. David Furley, "The Rainfall Example in *Physics* ii 8," in *Aristotle on Nature and Living Things: Philosophical and Historical Studies Presented to David M. Balme*, ed. A. Gotthelf (Pittsburgh: Mathesis Publications, 1985), 177–82.

32. *Physics* 198b34–199a3.

33. John M. Cooper claims that Aristotle's argument is based on the premise of the eternity of our world. See his "Hypothetical Necessity and Natural Teleology," in *Philosophical Issues in Aristotle's Biology*, ed. A. Gotthelf and J. Lennox (Cambridge: Cambridge University

Press, 1986), 243–74, esp. 246–53. Compare, however, *Meteorology* 347a5–6, where Aristotle himself says that the seasonal cycle of the rains merely "*wishes* to be perpetual in its order" (emphasis mine).

34. *Physics* 199a3–5.

35. Cf. *On the Parts of Animals* 676b16–677a19. This discussion (of the cause of bile) in *On the Parts of Animals* makes it clear that merely to be a necessary consequence of something good is not yet to be for the sake of something, at least not in Aristotle's view. Thus, for instance, the mere fact that death comes of necessity to living beings, and to intelligent living beings in particular, does not mean that it is for the sake of something (such as the existence of those beings).

36. *Physics* 199b34–200a32. See also 199b5–7.

37. Cf. *On the Parts of Animals* 646a8–647a3.

38. See above, pp. 41–42, and also consider n. 23.

39. See above, pp. 40–41. Cf. Maimonides, *The Guide of the Perplexed*, trans. Shlomo Pines (Chicago: University of Chicago Press, 1963), pt. 2, chap. 20, 313–14.

CHAPTER 3

On Continuity and Infinite Divisibility

In book six of the *Physics*, Aristotle confronts Zeno's notorious arguments that attempt to prove the impossibility of motion. Zeno had claimed, among other things, that any interval, however small, through which something might move contains infinitely many subdivisions and thus can not be traversed in a finite time. Aristotle's initial response to this argument is that time itself is also infinitely divisible, so that even a finite time contains the infinity that is needed for traversing the infinitely many divisions of length.[1] Later in the *Physics*, however, in the course of his own argument for the existence of an unmoved Prime Mover, Aristotle returns to this argument of Zeno and says that his earlier response to it, while being sufficient with a view to the questioner, was not sufficient with respect to the truth of the matter. For one could still ask, he says, about the time itself, namely, whether it is possible for its own infinitely many divisions to be gone through.[2] To this objection Aristotle replies that infinitely many divisions are only potentially present in time, as they are in any continuous quantity. And he adds that if one tries to make them actual, as for instance by counting the midpoint, and then the midpoints of the successive halves, one is no longer dealing with the original phenomenon: one has replaced—after only a single act of division—a continuous motion through a continuous interval of time with two distinct motions, and two distinct times, interrupted by a stop. Accordingly, Aristotle's revised answer to the question of whether it is possible to go through infinitely many stages, either of time or of length, is that it is not possible in the sense that such stages are actual, but that it is possible in the sense of their being potential. For this is what someone in continuous motion has done; he has gone through the infinitely many divisions that are potentially, but only potentially, present in any continuous interval.³

Precisely, however, if one is impressed by this revised version of Aristotle's answer to Zeno, one has to wonder why he does not present it when he first confronts Zeno's argument in book six. Why does he settle, even temporarily, for a mere refutation that accepts the misleading premise of his adversary and thus fails to bring the truth of the matter to light? This question is all the more necessary inasmuch as the premise in question, that everything in motion has gone through infinitely many stages, is not merely one that Aristotle has taken over from Zeno; quite independently of the argument with Zeno, he tries to establish this claim himself, and he argues for it at length, without any suggestion, in book six, that the infinitely many stages exist only potentially in a real motion. What is the reason for his strange delay before stating the fuller truth as he sees it? To try to answer this question, I propose to begin by examining the thematic accounts of the infinite and of the continuous that occur earlier in the *Physics*. I will then turn to the arguments in book six, which prove to be necessary in order to confirm that the infinite, understood as that which is divisible without limit, truly exists. By analyzing these arguments in detail, I will try to show that Aristotle's first response to Zeno, despite and even because of its misleading character, is a necessary part of the truly adequate response. In the course of my discussion, I will also try to show—and indeed this will be my main goal throughout—that the question of why Aristotle responds to Zeno in the way that he does leads to the heart of what the *Physics* has to teach us about nature.

Aristotle's preliminary justification for discussing the infinite in the *Physics* is that motion, which is clearly a central feature of all natural beings, is thought of as being continuous and that the infinite is first apparent in the continuous. Therefore, he adds, those who define the continuous often do so in terms of the infinite, by asserting that it is the infinitely divisible that is continuous.⁴ We shall see, however, that Aristotle's own definition of the continuous makes no mention of the infinite;⁵ and at the beginning of his thematic treatment of the infinite, he goes so far as to raise the question of whether it even exists. What he says there is that since the science of nature has to do with magnitudes, motion, and time, each of which is either infinite or finite, it would be appropriate for the student of nature to consider whether there is an infinite or not, and if there is, what it is.[6] Surprisingly, this passage speaks of being infinite as a simple alternative to being finite, and it thus seems to disregard Aristotle's earlier suggestion that the finite, if continuous, is infinite as well, in the sense of being infinitely divisible. Moreover, the opening sections of this discussion make almost no reference to the infinitely divisible.[7] Apart from a brief critique of the claim that there is an infinite principle of perceptible bodies that is itself separate from them all, Aristotle's initial focus is solely on the question of whether one or more kinds of the perceptible bodies are infinitely big, or extend infinitely far in all directions. And only after concluding, on the basis of arguments that can not concern us here, that there is no such infinite body does he turn to the consideration of the infinite by division.

The subject of the infinite by division is introduced in the following way. Aristotle has just argued that there is no actual infinite, and in particular no perceptible body that extends so far that it can never be traversed. He goes on to say, however, that many impossibilities clearly follow if there is simply no infinite: there will be a beginning and an ending of time, magnitudes as a class will not be divisible into magnitudes, and number will not be infinite. Since, then, neither the being of the infinite, as something actual, nor its unqualified non-being appears to be possible, Aristotle is led to assert that the infinite exists as a potential. Now in speaking of the infinite as a potential, he does not mean the kind of potential that could ever be actual in the same sense that a potential statue could at some time be an actual statue. Rather, he says, the infinite is like a day, or a series of games, both of which are what they are in actuality by one thing and then another always coming into being. Aristotle proceeds to divide the class of the infinite into two subclasses, one of which is seen in time and in the unending succession of human beings, and the other in the division of magnitudes, as in the successive division of a finite magnitude in a constant ratio. In the immediate sequel, however, he asserts that the infinite exists *in no other way* (emphasis mine) than potentially, by diminution.[8] He thus seems to deny the infinity of time, and of the succession of human beings, which he himself has just called to our attention. His reason for this reversal appears to be that these latter infinities exist only successively, and as the combination in thought of present being with a remembered or imagined past, whereas the infinite by division, or rather its finite substratum, is present as a whole.[9] So concerned is Aristotle with showing that the infinite has its being in something present as a whole that he will later derive the potential infinity of number from the successive addition of the (increasingly smaller) units that are brought into being by the division of a magnitude in a constant ratio.[10] These observations do not, to be sure, wholly resolve the question of the status of such infinities as that of time. But at all events, Aristotle does say repeatedly that the chief concern of the student of nature is with perceptible magnitude, or body; and if such magnitude does not extend beyond a definite limit, as he argues that it does not, it can be infinite only in the sense of being infinitely divisible.[11]

According to Aristotle's account, perceptible magnitudes are not even potentially infinite in the direction of increase. For however many times one adds proportionally smaller amounts to any magnitude, their sum will never surpass a definite limit. And if one keeps adding equal amounts, their sum will eventually surpass the size of the heaven, which he claims that no perceptible magnitude can do. On the basis, then, of this ultimate limit to the increase of size, together with the principle that "it is possible to be as big in actuality as it is possible to be in potentiality," Aristotle concludes that magnitude or body is not infinitely big even in potentiality.[12] In the case of number, however, he allows for the infinite as a potential, since despite the fact that no number is infinite in actuality, one can always, at least in thought, exceed every determinate number by adding more units to it. Yet one wonders whether this allowance for the infinite as potential in the realm of number is compatible with the claim that it is possible to be as big in actuality as it is possible to be in potentiality. If every actual number is finite, as it is, it would seem to follow that there is also not even a potential infinite in number.[13] To be sure, there might perhaps be such an infinite if there were some being or group of beings that had the time and the ability to keep on counting new numbers forever, for in that case the potentially infinite could also have the kind of actuality that consists of one thing following another without end; but on the reasonable assumption that there are no such beings, it remains difficult to see how number is even potentially infinite.[14]

For the same reasons, moreover, that one must wonder about Aristotle's claim that there is a potential infinite in number, one must also wonder about the infinite divisibility of magnitude. And this question about the infinite divisibility of magnitude becomes even more pressing if we limit our attention to perceptible, as distinct from mathematical, magnitude. For how do we know what happens to perceptible magnitudes after they have become too small to be actually perceived? Even if they could always be thought of as being further divided, what is possible in thought is not necessarily possible in fact, as Aristotle himself stresses in his argument against an actual infinite.[15] Aristotle's claim regarding the infinite divisibility of magnitude is also in tension with another aspect of his account. For he speaks of the infinite as the material of the completeness of magnitude, a material that is a whole not in itself, but only by virtue of the form that encompasses it and gives it definition. And since it is only by virtue of its form that anything can be known as what it is, he adds that the infinite is unknowable insofar as it is infinite.[16] But if the infinite, as such, is unknowable, how can we claim to know that the infinity of a magnitude is its divisibility without limit into other divisible magnitudes? Aristotle's account of the infinite, then, has not established that there is such an infinite as he has described, especially not in the realm that is of most concern to the student of nature, the realm of perceptible magnitudes. And he himself is aware of this deficiency, as we can see from a remark near the end of his account. He says there that for the time being he is merely using the (terms) infinite in magnitude, infinite in motion, and infinite in time, but that later he will say what each of these is and why every magnitude is divisible into magnitudes.[17] By thus implicitly denying that he has told us what the infinite is, he acknowledges the legitimacy of our questions as to whether it exists. For the bulk of his account has been a clear statement of what he means by the infinite, and in particular by the infinite in magnitude. But as he argues in the *Posterior Analytics*, one can not know *what* something is, as distinct from what the meaning is of a term, unless one knows *that* it is.[18] Thus, it makes sense that at least one of the reasons for his admission that he has not yet said what the infinite is, is an awareness on his part that he has not yet shown that it exists.

Aristotle has said, we recall, that the infinite is first apparent as the infinitely divisible in what is continuous.[19] Accordingly, he gives his own definition of the continuous before trying, in book six, to make good on his promise to say what the infinite is. He defines the continuous as a subclass of the contiguous, which itself is a subclass of the successive. Something is successive, he says, if nothing of the same kind comes between it and its predecessor. He then says that it is contiguous if, in addition to being successive to its predecessor, it touches it. According to this account, then, a line, an arithmetical unit, or a house can be successive to another, but only between the lines or the houses can there be contiguity. Finally, Aristotle says that the continuous is found when two contiguous elements have become one, held together as a unity by virtue of the fact that the limit of each, where they are touching, has itself become one and the same. He adds that a continuous quantity could not be held together as one if there were still two extremities at a juncture within it. Yet this claim, though it makes sense in itself, is hard to square with the fact that he first spoke of the continuous as a subclass of the contiguous, since things that are contiguous are themselves different, and thus would seem to have different extremities as well. Now it is true that in his examples of the various ways in which things become continuous, Aristotle implies a somewhat looser definition of continuity, and this looser definition would indeed be compatible with being contiguous. For he says that things can become continuous by being nailed together, by being glued together, by touching, or by natural adhesion; and the extremities of two boards, for instance, that have been nailed together in constructing a house do not become strictly speaking one, however irrelevant to us the difference between them might be. Still, when Aristotle first spoke of the continuous as a subclass of the contiguous, he was apparently referring to it as such, and not merely to what is continuous in an extended sense of the word. He confirms this impression in the sequel, where he argues that things must necessarily touch, that is, they must be contiguous, if they are continuous. And his use of tenses in this context makes it clear that he is presenting their touching not merely as a precursor to their becoming continuous, but also as something that persists after they have become so.[20] Later, we shall have to try to make sense of this puzzling relation between the definitions of the continuous and the contiguous. But for now, we return to the fact that Aristotle's thematic account of the continuous, according to which things are continuous if the extremities between them are one, has made no mention of the infinite. It has, however, laid the basis for his attempt to show that the continuous is indeed infinitely divisible and that the infinite, therefore, truly exists.

Aristotle's argument that continuous quantities are infinitely divisible is the primary theme of book six of the *Physics*. This argument begins negatively, with a proof that the continuous is not made up of indivisibles—on the assumption, it seems, that if it is not, then neither is it divisible into indivisibles, and that the only alternative is to be infinitely divisible. To prove that the continuous is not made up of indivisibles, Aristotle relies on his definition of the continuous as that subclass of the contiguous in which the extremities have become one. He also limits what he means by the indivisible to that which has no parts or inner differences, even in thought; and he thus does not consider, for instance, the hypothesis of an extended body that remains indivisible by virtue of its impassivity alone.[21] His primary argument is that since an extremity is different from that whose extremity it is, an indivisible can have no extremity to be one with that of another, or even to be in the same place with it.[22] He adds a number of similar arguments that the continuous can not be made up of, or divided into, indivisibles, and thus it seems that it must be infinitely divisible, and that the infinitely divisible must exist.

Someone could still object, however, that continuity itself is only an illusion, and that what we call "continuous" quantities are in fact agglomerates of indivisibles, if not of indivisibles in Aristotle's narrow sense of the word. To meet this implicit objection, Aristotle turns to the specific cases of magnitude, motion, and time, which are normally thought of as being continuous, and he tries to show that infinite divisibility applies to them. To do this, he begins by showing that magnitude, motion, and time are so related that if any one of them is made up of indivisibles, then so are the others. In particular, he argues at greatest length that if magnitudes are made up of indivisible intervals, things would have to have completed their motion through each of these intervals without previously being in the process of moving; for any such process would be the traversing, and hence the dividing, of the allegedly indivisible interval. Nothing, therefore, would ever be moving, but only in the state of having moved, and motion would be made up of indivisible "moves" (or "quantum leaps," as we might call them today). Now Aristotle evidently regards this as an absurd consequence, and to emphasize its absurdity, he couples it with the hypothesis that everything is necessarily either moving or at rest, from which it follows that a body in motion through indivisible intervals would be continuously at rest at the same time.[23] Since, then, the premise that magnitude is made up of indivisibles has led to such apparently absurd results, it seems necessary that it, along with motion and time, must be infinitely divisible.

Aristotle turns next to another argument for infinite divisibility in magnitude and time. He claims that it has been shown, in effect, that every magnitude is divisible into magnitudes, or in other words that magnitude is infinitely divisible, and he proceeds to argue on this basis that time is infinitely divisible as well. For since, he says, every magnitude is divisible into magnitudes, in whatever time it takes for something in motion to traverse a magnitude, however small, something moving more slowly will have traversed a still smaller one, which smaller magnitude was in turn traversed by the faster in still less time, and so on.[24] After concluding in this manner that time is infinitely divisible—or rather continuous, as he calls the infinitely divisible in this context—Aristotle notes that the argument, in which magnitudes and times are repeatedly divided in turn, has also made clear the continuity of magnitude. Yet since the continuity or infinite divisibility of magnitude was presupposed in the first place, in order that something moving more slowly than another could traverse a smaller magnitude in however small an interval of time, this additional conclusion makes the whole argument seem circular.[25] And the impression of circularity is reinforced in the sequel, where Aristotle gives what he calls one of the customary arguments that if time is continuous, then so is magnitude; for this new, hypothetical argument suggests that the original assertion that every magnitude is divisible into magnitudes is still in doubt. And perhaps there is good reason for doubt. For to the extent that these arguments for infinite divisibility are not merely mathematical, they have relied uncritically on common opinion and on extrapolation from what appears to our senses. We can not accept them as proof of infinite divisibility in nature—even if we disregard the question of how such an infinite could become actual—unless we trust that the magnitudes, or else the motions and times, that we perceive as continuous would preserve this character when divided into imperceptibly small parts.

Aristotle openly acknowledges a bit later in book six that he has not yet made the case for infinite divisibility. Just prior to this acknowledgment, he has broadened the discussion by introducing the topic of that which moves or changes, where change is understood to include alteration (or change in quality) and coming into being (or change in kind) as well as change in size and in place. He has argued that the changing being is necessarily divisible, and divisible into as many divisions as are its motion, the time of its motion, the magnitude through which it moves, and the like, and he asserts that its divisibility is the chief ground of that of the others. In his words, "that they are all divided and that they are infinite follows above all from that which changes. For the divisible and the infinite are inherent immediately in that which changes."[26] Now in this formulation the term *infinite* appears to mean infinitely divisible; for though Aristotle has also spoken of the infinite by addition, or "in its extremities,"[27] he has emphatically denied that any body or changeable being is infinite in this sense. Thus, he seems to distinguish the mere divisibility from the infinite divisibility of that which changes. And here is where he acknowledges that the question of infinite divisibility is still in doubt. For despite his earlier arguments, and despite the fact that he has just spoken of "the infinite," along with "the divisible," as being "immediately" inherent in that which changes, he goes on to say that "the divisible has been shown earlier, but the infinite will be clear in what follows."[28]

To understand how Aristotle can claim to have made the case for divisibility, without yet having shown it to be infinite, let us look back to his argument that everything that changes must be divisible. He begins by noting that all change is from something to something. He then says that what is changing is no longer changing when it is in that to which its change was proceeding, and that it is not yet changing when it and all its parts are in that from which the change began. Accordingly, he reasons, while it is changing, one part of it must be in the "from which" of change and another part in the "to which," since it can not be in both or in neither. He therefore concludes that everything that changes must be divisible.[29] One could raise as an objection to this argument, as I have presented it, that when something is changing from white, say, to black, it is of neither color, not even in part, but rather wholly in between. But Aristotle meets this objection by specifying that when he speaks of that to which a change proceeds, he means the *first* terminus to which it changes, even if the change should continue further. Thus, for instance, if grey were the first color into which something changes during the course of a larger change from white to black, his argument would apply to the change between white and grey. At any time during this change, one part of the changing surface would have to be white, and another part would have to be grey. Now if we trust this argument for the divisibility of everything that changes, we must presumably accept infinite divisibility as well, on the grounds that the analysis of the changing being as a whole can also be applied separately to each of its changing parts, and to the parts of those parts, and so on.[30] But Aristotle himself has given us reason to doubt whether his argument applies to everything that changes. Earlier in the *Physics*, he had criticized Melissus for failing to recognize that there are changes in which a being is transformed simultaneously in all its parts.[31] He will later mention freezing as an example of such change, in which it is false—at least if the body that changes is relatively small—to say that some parts are altered before others.[32] This kind of change, which Aristotle speaks of as being frequent, and which may even be the norm in the alterations of very small bodies, is clearly inconsistent with his argument for the divisibility of everything that changes. To be sure, the conclusion of the argument is not itself called into question, for as the commentator Themistius has well observed, something that changes in all its parts at the same time must necessarily have parts and thus be divisible.[33] But since we do not yet know the basis for Aristotle's account of these changes, we can not yet be certain that the divisibility which it requires is divisibility into infinitely many parts. Consequently, and in view of the difficulties we have seen in the earlier arguments as well, we need not be surprised at his subsequent admission that "the infinite," or infinite divisibility, has not yet been shown to exist.

Aristotle has promised, we recall, that the existence of the infinite would become clear in what follows. And in the immediate sequel, he at least continues with this theme. He argues, for instance, that there can be no first interval of time in which a change occurred or was occurring.[34] Yet this argument explicitly assumes, without even an attempt at proof, that there are infinitely many divisions of time. And it leads to another difficulty as well. For on the very next page, Aristotle seems to contradict his own conclusion by asserting that there *is* a first interval of time in which something changes. He understands this interval as being first in relation to any number of larger ones, in which the change is also said to take place, but only in the derivative sense that these larger intervals contain the first or primary one. To be sure, he then argues that what distinguishes this first interval from the others is that the changing being must have been changing in any part of it.[35] But he does not for that reason retract his claim that it is first, and thus there remains at least a nominal contradiction with his previous assertion that there is no such interval. Now from the context, it appears that Aristotle's reason for continuing to speak of a particular interval as the first in which something changes is that he thinks there is a definite instant at which a change is first complete; the first interval of time would thus be the one whose end point is that culminating instant.[36] This suggestion, however, does not wholly resolve the difficulty, as we see from Aristotle's subsequent argument, in which he elaborates upon his claim that no interval of change can be first. What he argues is that everything that is changing must have changed, or completed a change, previously, and indeed that it must have completed infinitely many changes in the infinitely many intervals into which the time of its changing is divided.[37] But if this is true, the so-called first interval of time in which something is changing would be the first interval only for an arbitrarily selected change and would in fact be preceded by infinitely many others. Why, then, does Aristotle persist in speaking of a first interval of time in which something changes?[38]

To try to answer this question, and in the hope of clarifying the still unresolved matter of the infinite, let us turn to the next in this series of arguments. Here Aristotle reverses the order of terms from the preceding argument and tries to show that whatever has changed, or completed a change, must have been changing before. In order to make this case, he begins by pointing out that everything that has changed must have done so in time, as distinct from an atomic instant. He then adds that since all time is infinitely divisible, what has changed in time must have changed already in each of its infinitely many subsections. And on this basis, he concludes that what has changed must have been changing, or in the process of changing, before.[39] Aristotle admits at once, however, that this argument is not equally clear in the case of every kind of change. For he holds that the intervals between contrary qualities, as well as the interval between the contradictories of non-being and being, are not continuous, or infinitely divisible, as are intervals of magnitude. Thus, although he can speak of infinitely many subdivisions of a magnitude as having been traversed in the infinitely many time-intervals of a motion through it, his denial that there are infinitely many qualities between two extremes raises doubts about his assertion that what is changing between these extremes must have already completed infinitely many changes.[40] For how can there be a completed change in quality without some definite quality to be its terminus? That there are only finitely many qualities between any two extremes was already implicit in Aristotle's earlier reference to a first color into which something changes. And though he does not give an argument in the *Physics* in support of this general assertion, there is no one, however subtle, who even imagines that he can distinguish infinitely many colors or sounds.[41] Accordingly, the question of how something that is changing in quality can have completed infinitely many changes does indeed arise. Now earlier, Aristotle had suggested a way out of this difficulty by claiming that the being that changes in quality is an infinitely divisible magnitude, so that even its changing to a first new quality would be preceded by the completed changes of its infinitely many parts.[42] But as we have noted, this suggestion fails to take into account those changes in which according to Aristotle himself the changing being is changing simultaneously in all its parts. And at

On Continuity and Infinite Divisibility   65

all events, he does not have recourse to this suggestion here. Instead, he says, in order to make the case that even what has changed through a discontinuous interval must have been changing before, "we will take the time in which it has changed, and we will again say the same things."[43] This remark seems at first to mean that the infinite divisibility of time suffices by itself to guarantee that any being that has completed a change has already completed infinitely many others.[44] But on this view, Aristotle would be merely avoiding the question, and so let me suggest a more adequate interpretation of his remark. For in order to show, as he is trying to do here, that whatever has changed was previously in the process of changing, was there ever any need to argue that time is infinitely divisible, or that the being that has completed a change has already completed infinitely many others? Moreover, even if these claims could both be shown to be true, would they suffice for showing that the being in question was previously in the process of changing? For one could still ask, regarding each of the infinitely many completed changes, whether it was preceded by a process of changing. And if the being that has changed was not changing in the time between each of these completed changes, why should it be changing during the whole time? Accordingly, I suggest that Aristotle's genuine argument that it was indeed changing consists entirely in his reminding us that it has changed in an interval of time, and not in an atomic instant. Once he has said this, there is no more for him to do except to say the same things again until they sink in. For we speak of something as having changed in time precisely because it has been changing throughout that interval. And if the question still persists of how one knows that it was changing before it completed its change, the simple answer is that we can perceive it. We can perceive this process of changing even when, as in the case of a puddle that freezes, something changes simultaneously in all its parts, and between qualities that have no intermediate. And since in such cases, at least, the completed change is not preceded by other completed changes, it also makes sense to speak of a first or primary time during which something is changing.[45]

    This last suggestion about what it means for there to be a first time in which something is changing can also help us to confirm that the infinite exists. For during the time in which something is changing through an indivisible interval of quality, as for instance between a liquid and a solid state, it has neither its initial quality nor its final one, at least not simply. And though Aristotle does assert that a being's state is that in which most, or the most important, of its parts are found, even if it is not wholly in that state, this suggestion can not help us to assign a quality to something that is changing simultaneously in all its parts.[46] At least in such cases, then, the changing being must have a kind of infinity, or indeterminateness, with respect to quality.[47] To be sure, this infinity is not the same as the infinite divisibility that Aristotle has been discussing at such great length. And yet he has explicitly called our attention to this other kind of infinity as well. For early in the *Physics*, in his criticism of Anaxagoras, he distinguished between the infinite with respect to form (including perceptible form or quality) and the infinite with respect to multitude or magnitude.[48] And though his thematic discussion has been almost wholly concerned with the latter of these infinites, his explicit reference to the former kind makes it thinkable, at least, that this is what he had primarily in mind when he said it would become clear that the infinite exists.

These suggestions about the infinite, and about the first time in which something changes, have both involved a reliance on our perception. This reliance is not identical with the kind that I criticized earlier, for to perceive the process of changing between discontinuous qualities or forms does not depend on the extrapolation that underlies the claim that magnitude and motion are infinitely divisible. Still, it is reliance on perception, and one might wish for something deeper and more certain. One might still suspect that, appearances to the contrary, motion is made up of imperceptible quantum leaps or "moves." Indeed, Aristotle mentions this alternative again at the end of book six, after he has spelled out in detail his first response to Zeno;[49] and though he says it has been shown to be impossible, one might doubt the force of his arguments. For if we look back at these arguments, we will notice that Aristotle has distinguished between *motion* (the noun) and *to be moving* (the progressive infinitive), and he has left it as a mere hypothesis that something must be moving whenever motion is present.[50] We may thus still wonder whether what we call "motion" is not something of an illusion, and that natural beings must ever be simply at rest, not even with respect to the qualities to which they alter.

We can gain further insight into Aristotle's understanding of motion and rest, and of their relation to our perception, by comparing his two responses to another extreme challenge to perception, which he turns to next. According to this view, even if both motion and rest exist, nothing is both sometimes in motion and also sometimes at rest. Now Aristotle's first response to this claim is the simple assertion that "we see the aforementioned changes coming about in the same things."[57] It is not quite clear, however, what he means by "the aforementioned changes"; perhaps he is referring to motion and to coming to rest, which he has distinguished from motion in the discussions in book six.[58] But at all events, what is most important for our purposes is that he does not say that we see *being* at rest, at least not if rest is understood as a state that excludes change. Accordingly, he supplements this appeal to perception with the assertion that this new claim is also at odds with what is evident, with the universally held belief, for instance, that "to be in motion is to become something [i.e., something fixed or at rest] and [for something] to perish."[59] Nevertheless, when he recapitulates his response to this claim, he does say that we see things at rest, and this time he does not go on to offer any other kind of evidence than that of perception. More precisely, he says that there is "one sufficient assurance" [ἱκανὴ μία πίστις] against all three of the claims that we have been considering, namely, that "we see some things sometimes in motion and sometimes at rest."[60] Here, then, in his first explicit and unequivocal statement that we see things at rest, Aristotle also says for the first time that our perception is a *sufficient* assurance against these various claims that would challenge its evidence. And it is perception alone that he speaks of as such an assurance. Now perception can be such an assurance only on the condition that we do not ask it to tell us what it cannot tell, as for instance what there might or might not be beyond its range of discrimination. But if we accept this condition, it is indeed evident that we do perceive things at rest, in the sense that there are sometimes no noticeable changes in a being, or at least none with respect to the predominance of a particular character or quality. If we did not perceive such rest, there could not even be a world of discrete beings to investigate. Accordingly, when Aristotle says here that what we see is a sufficient assurance against these extreme challenges to our common beliefs about the world, he must also mean that the beings as we perceive them, and not, for instance, any imperceptible conditions of their being, are the true objects of the study of nature. And it is in this sense, I think, that we must understand his well-known remark in *On the Heaven* that "the culmination . . . of natural [science] is that which appears, always authoritatively, in accordance with perception."[61] Aristotle is not merely saying, as a modern scientist might, that we must test our hypotheses about nature in terms of the consistency of their predictions with what we actually perceive. He is also saying, more importantly, that the manifest character of natural beings as they are given to us in perception must be understood as their true character, and not merely as our inlet to a deeper nature.[62]

Aristotle's basis for relying on our perception of beings in motion and at rest is also applicable to our perception that motion is continuous, rather than the sum of indivisible and imperceptible leaps. For even if it were the case that the motions we perceive were somehow conditioned by such leaps, these would not be constituents of the motions themselves, whose continuity is immediately evident to us. But the continuity that we are thus aware of is not, nor does it necessarily imply, infinite divisibility.[63] Rather, it is the perceived oneness of both the character and the time of a single being's motion from beginning to end.[64] Similarly, the continuity that we are aware of in the case of magnitudes and in that of time is their perceived unity, or their lack of perceptible breaks. And since this perceptible continuity is the only kind that we can be certain of in nature, we can understand why Aristotle must go on to correct his first response to Zeno, which had depended on magnitude and time being infinitely divisible.

We recall that Aristotle's revised response to Zeno is that despite the infinitely many potential divisions in continuous quantities, they are undivided in actuality. And we are now prepared to see some further implications of this response. For according to Aristotle's precise argument, the actual positing of potential divisions, if only by counting the halves in a continuous whole, is destructive of continuity, since it compels us to treat a single midpoint as two, as the end of the first division and as the beginning of the second.⁶⁵ But if this is true, then the definition of the continuous as that in which the extremities of contiguous quantities have become one, the definition, that is, from which Aristotle argued for its infinite divisibility, must also be revised. If a single midpoint determined in actuality is necessarily also two—at least in our perspective, which is the perspective in which the natural is revealed as what it is—then a continuous quantity must have no actual midpoints, and no actual parts that touch at an actual extremity. It is because none of its potential midpoints is present in actuality that the continuous is truly different from the contiguous.⁶⁶ And this is also why a continuously moving body is never actually at any of these midpoints and why it never traverses any actual interval that they mark off.⁶⁷ But the question must now arise, in the light of our earlier discussions, of how these midpoints and these intervals can be infinite even in potentiality. We recall Aristotle's assertion that it is possible to be as great in actuality as it is possible to be in potentiality, and we have seen good reasons to deny that the number of midpoints or intervals can even become actually infinite, to say nothing of being so all at once.⁶⁸ Accordingly, I suggest that when Aristotle says here that a body in motion traverses infinitely many divisions in potentiality, he does not mean that there are even potentially more of these divisions than every definite number, but only that they are indeterminately many. He is saying no more than that the path of this motion might be divided anywhere, and that the number of points where division is possible can not be determined with exactness.⁶⁹

My interpretation, in this chapter, implies that Aristotle has not been in earnest in many of his arguments about the infinite. If I am correct, he is well aware that it is neither possible nor necessary to show that the continuous, especially as it exists in natural beings, is infinitely divisible, in the sense that every division could always be followed by another. But if this is true, we are left with the question of why he would argue at such length for a position that he does not seriously intend to show. This question, however, is merely a more specific version of the general question regarding the manner in which he presented his account of nature. Thus, rather than my responding to this particular question now, I will return to it in my concluding discussion of Aristotle's manner of writing.

NOTES

1. *Physics* 233a21–31; cf. 239b9–30.
2. *Physics* 263a4–23; cf. 239b28–29.
3. *Physics* 263a23–b9.
4. *Physics* 200b16–20.
5. *Physics* 227a10–17.
6. *Physics* 202b30–36.
7. The exceptions are *Physics* 203b17 and 204a6–7.
8. *Physics* 206b12–13. Aristotle does proceed to speak of the infinite as existing also in actuality, in the sense that a day or a series of games does. By this, he presumably means that the potentially infinite, finite magnitude can also be infinite in actuality, in a sense, during the time when the successive division is actually being carried out.
9. *Physics* 206a33–b3, 208a20–21; 207a21–24, 208a1–2.
10. *Physics* 207b7–15.
11. *Physics* 204a1–2, 204a34–b4, 205a7–9.
12. *Physics* 207b17–21; cf. 206b16–26; and see also 206b12–14. Aristotle does not give any argument in support of the principle that I have quoted in the text, since it belongs to the very meaning of the term *potentiality* that whatever something is (or can be) in potentiality it can also be in actuality, at least in some sense of the word *actuality*. See *Metaphysics* 1047a24–26, b3–6.
13. Cf. Simplicius, *In Aristotelis Physicorum Libros Quattuor Priores Commentaria*, in *Commentaria in Aristotelem Graeca*, vol. 9, ed. H. Diels (Berlin, 1882), 498.28–31, 508.3–509.20. See also *Physics* 207b27–31.
14. Aristotle does, to be sure, assert that the human race is eternal. But he also argues that periodic cataclysms and other such disasters destroy the continuity of peoples and of civilizations. For this reason, then, if for no other, the human race can not in his view be the agent of a perpetual count. See *On the Generation of Animals* 731b35–732a1; *Meteorologica* 339b27–30, 351a19–353a28; *Politics* 1329b25–27.

Since Aristotle regards time as number, and number as that which can be counted, his denial that the human race can maintain a perpetual count helps to explain his implicit retraction of the claim that time is infinite. See pp. 55–56, and compare *Physics* 223a16–29. See also Themistius, *In Aristotelis Physica Paraphrasis*, in *Commentaria in Aristotelem Graeca*, vol. 5, pt. 2, ed. H. Schenkl (Berlin, 1900), 101.9–13.

15. *Physics* 208a14–19; *On the Heaven* 299a11–17; cf. Simplicius, *In Libros Quattuor Priores Commentaria*, 493.33–494.11.
16. *Physics* 207a21–32.

17. *Physics* 207b25–27.
18. *Posterior Analytics* 93a14–27; cf. 89b31–35, 92b4–8; *Physics* 202b34–36.
19. *Physics* 200b17–20; cf. 207b35–208a2.
20. *Physics* 226b34–227b2; cf. *Metaphysics* 1015b36–1016a17.
21. Aristotle argues against this more "physical" version of atomism, and in particular against the doctrines of Democritus and Leucippus, in *On Coming into Being and Perishing* 325b34–326b6.
22. *Physics* 231a21–29. See also Simplicius, *In Aristotelis Physicorum Libros Quattuor Posteriores Commentaria*, in *Commentaria in Aristotelem Graeca*, vol. 10, ed. H. Diels (Berlin, 1895), 925.5–24. This is also translated by David Konstan, in *Simplicius: On Aristotle Physics 6* (London: Duckworth, 1989), 17.
23. *Physics* 231b18–232a17.
24. *Physics* 232a23–233a10. Just before this, Aristotle had made the simpler argument that if every length (and motion) is divisible, and if something moving uniformly goes through a smaller length in less time, then time is (infinitely) divisible as well. But in that earlier argument, whose purpose was merely to show that if time is indivisible then so is length, the premise that every magnitude is infinitely divisible was still treated as a hypothesis, and not yet asserted as a fact. *Physics* 232a18–22; cf. Alexander as reported by Simplicius, *In Libros Quattuor Posteriores Commentaria*, 937.25–28 (also in Konstan, *Simplicius*, 31).
25. Cf. Simplicius, *In Libros Quattuor Posteriores Commentaria*, 944.1–15ff. (also in Konstan, *Simplicius*, 37).
26. *Physics* 235b1–4.
27. Cf. *Physics* 233a16–21.
28. *Physics* 235b4–5.
29. *Physics* 234b10–20.
30. Cf. *Physics* 234b21–235a10, 236a27–35.
31. *Physics* 186a10–16.
32. *Physics* 253b23–26. See also *On Sense and Sensible Objects* 446b28–447a6.
33. Themistius, *In Aristotelis Physica Paraphrasis*, 191.30–192.22. Cf. Simplicius, *In Libros Quattuor Posteriores Commentaria*, 966.15–969.24 (also in Konstan, *Simplicius*, 62–66).
34. *Physics* 236a14–27.
35. *Physics* 236b19–32.
36. *Physics* 235b30–236a13; cf. 236b34–35.
37. *Physics* 236b32–237a17.
38. Cf. *Physics* 238b31–239a10.
39. *Physics* 237a17–28.

40. *Physics* 237a28–b2.

41. Cf. *Physics* 236b1–18; *On Sense and Sensible Objects* 445b20–446a20.

42. *Physics* 236a27–b8.

43. *Physics* 237a35–b3.

44. Cf. Simplicius, *In Libros Quattuor Posteriores Commentaria*, 996.6–12 (also in Konstan, *Simplicius*, 95–96); *Aristotle's Physics: A Revised Text with Introduction and Commentary*, ed. W. D. Ross (Oxford: Clarendon Press, 1936), 652.

45. Support for this interpretation of Aristotle's argument can be found in the immediate sequel, where he speaks of coming into being, rather than change, and asserts that what has come into being must necessarily be coming into being previously, and that what is coming into being must likewise have come into being (*Physics* 237b9–22). For not only does he add that what has already come into being need not be identical with what is coming into being—but might, for instance, be one of its parts—but he also explicitly limits at least the second half of his claim to the becoming of that which is divisible and continuous. And one can make sense of this limitation on the basis of what I have argued in the text: although a color, for instance, must have been coming into being before having come into being, it may be coming into being without it, or indeed any color, having come into being already. (A comparison of 237b15 [συνεχεῖ γε ὄντι] with 235b3–4 [and a15–18] shows that in this argument, unlike the earlier one, Aristotle is thinking of qualities and forms as among the possible subjects of coming into being.) Consider also Themistius, *In Aristotelis Physica Paraphrasis*, 197.16–25, along with Spengel's suggested emendation to line 24.

46. *Physics* 240a19–29.

47. In the case of change between contradictories, where the changing being must necessarily be either in one state or the other at all times, its infinity or indeterminateness would coexist with its being within the range that defines (first) one state and (then) the other, rather than being an indeterminateness of quality between two contraries. For an instance of such change between contradictories, see p. 68. David Bostock's argument that a change between contradictories cannot take time disregards the fact that the beginning of the new state (or of the new being)—whether or not it occurs successively among the parts of the whole being—is the terminus of a process of change during which the being is still within the range of its earlier state (or during which the new being is not yet in existence). See David Bostock, "Aristotle on Continuity in *Physics* VI," in *Aristotle's* Physics: *A Collection of Essays*, ed. Lindsay Judson (Oxford: Clarendon Press, 1991), 197ff.

48. *Physics* 187b7–11; cf. 207a24–32. See also *On the Soul* 424a17–24.
49. *Physics* 240b30–241a6; cf. 241a23–26.
50. *Physics* 231b25–27; cf. 235a15–16, 235a25–34.
51. *Physics* 253a32–b6; cf. 184b25–185a17.
52. *Physics* 253b6–26.
53. *Physics* 253b26–27.
54. Cf. *Aristotle's Physics*, translated with commentaries and glossary by Hippocrates Apostle (Grinnell, Iowa: Peripatetic Press, 1980), 316. Compare *Metaphysics* 1010a7–19.
55. *Categories* 12a4–10.
56. *Physics* 253b28–30.
57. *Physics* 254a3–7.
58. Cf. *Physics* 238b23–239a10.
59. *Physics* 254a8–15.
60. *Physics* 254a30–b4; cf. *On Coming into Being and Perishing* 327a14–22.
61. *On the Heaven* 306a16–17.
62. Cf. *On the Heaven* 306a9–11; *On the Soul* 425b26–426a26.
63. Wolfgang Wieland rightly emphasizes that Aristotle differs from modern mathematicians by limiting the focus of his inquiry into the continuum to the continuum as it is evident to us through perception. He also suggests, however—and inconsistently, in my view—that Aristotle understands this continuum as being infinitely divisible. Wolfgang Wieland, *Die aristotelische Physik: Untersuchungen über die Grundlegung der Naturwissenschaft und die sprachlichen Bedingungen der Prinzipienforschung bei Aristoteles* 2nd ed. (Göttingen: Vandenhoeck and Ruprecht, 1970), 278–89; contrast 305–7 with 326.
64. Cf. *Physics* 228a20–b11.
65. *Physics* 263a23–b9.
66. Aristotle's earlier hesitation as to whether to define the continuous as something distinct from the contiguous or else as a subclass of it stems from the fact that he was not yet ready to assert that a continuous quantity has no midpoints in actuality.
67. Compare *Physics* 262a28–31 with 262b28–263a1.
68. See again *Physics* 207b17–18. Compare Simplicius, *In Libros Quattuor Posteriores Commentaria*, 1293.3–5, where he says that defining the continuous as the infinitely divisible is like defining a bull as that which is able to become a bee. (At the end of the *Georgics* [bk. 4, vv. 281–558], Virgil tells the story that the Egyptians have learned to slaughter bullocks and prepare their carcasses in such a way that bees are brought forth from the rotting flesh. He traces the origin of this skill

to the shepherd Aristaeus, who was told by his mother, the nymph Cyrene, that he must perform a ritual slaughter of bulls and heifers in order to appease the Nymphs who had killed his bees. According to the story, nine days after Aristaeus had sacrificed these animals, a swarm of bees came forth miraculously from their entrails. See Liddell and Scott, *A Greek-English Lexicon*, 9th ed. [Oxford: Clarendon Press, 1940], s.v. "βουγονής." Also compare Judges 14.5–9.)

69. Cf. *On Coming into Being and Perishing*, 316a14–317a17.

CHAPTER 4

The Question of Place

The discussion of place in book four of the *Physics* contains one of its fullest statements regarding the manner in which one should conduct a scientific inquiry. This statement helps to elaborate on remarks in the opening chapter of the whole work, where Aristotle had said that while science itself must begin from the principles of the beings under investigation, the natural pathway to discover those principles is to proceed from what is better known and clearer to us toward what is clearer and better known by nature. He had mentioned there, as examples of what is better known to us, the beings as they appear to sense perception; and he had argued that our familiarity with these perceptible wholes is the natural beginning point for inquiry into their intrinsically more knowable principles. In the discussion of place, however, he makes it explicit that the pathway to science must also include some reliance on our beliefs about the matters in question, beliefs that we all bring with us prior to reflection, and that incorporate much of our initial grasp of the world. Now this reliance on our beliefs is not unproblematic, at least not in the case of our beliefs about place, since some of them give rise to such perplexities that the very existence of place must be called into question.[1] And yet Aristotle still asks us to assume that those things that are truly believed to belong to place do belong to it in fact.[2] Later, we shall examine in detail the attributes that he has in mind. But for now it is more important to note that he does not ask for our unqualified trust in the assumption that place possesses them. What he suggests, rather, is that by beginning from the mere postulate that it does (i.e., a postulate regarding the *kind* of thing place is), we can hope to arrive at genuine insight into *what* it is, insight on the basis of which we could finally be certain, not only that it exists, but also that it possesses these very attributes. As he states the matter in his own words, "On the basis of these assumptions [namely, that place possesses the attributes in question], one must contemplate what remains. And one should try to make the inquiry in such a way that its 'what it is' will be given, in order that the perplexities might be resolved, that the things believed to belong to place will indeed belong to it, and also that the cause of our discomfort and of the perplexities regarding it will be manifest. For this is the most beautiful way in which each thing could be brought to light."[3] Aristotle does not assure us, however, at least not in this passage, that so thoroughly satisfactory an answer to our questions about place can be attained. All he has said is that "one should try" to inquire in such a manner that the truth about place would be brought to light in this "most beautiful" way. But can all this be done? The answer to this question would shed considerable light, moreover, on the more general question of how Aristotle understands the pathway from our initial beliefs about the world to a genuine knowledge of its true character. And it is with these questions in mind that I propose to try to interpret his discussion of place.

Aristotle begins his discussion by saying that the student of nature must know about place, that is to say, whether it exists or not, in what manner it exists, and what it is.[4] This topic, he says, raises many perplexities, for despite the universal assumption that the things that are, are somewhere, or in some place, all the attributes that are thought to belong to place do not appear, upon consideration, to belong to any one and the same thing. Before turning to these perplexities, however, Aristotle presents the arguments that are thought to make it clear that place exists. The first of these is based on the replacement of bodies by one another. Water, for instance, sometimes departs from where it was, as from a vessel, and is replaced by air; or perhaps some other body comes to occupy this same place, which is thought, accordingly, to be something different from all the bodies that move into it or from it. A second argument, which seems to show that place, in addition to being something, has some power, is based on the motion of each of the natural elements to its own place. For the natural motion of light bodies upward and of heavy ones downward seems to show that at least these two places, the up and the down, are distinguished not merely by their position in relation to some arbitrary observer, but also in nature itself, by their having different powers or capacities. Aristotle goes on to claim that those who say that the void exists also imply the existence of place, since the void would be place deprived of body. And from these arguments, he says, one might come to assume that place is something apart from or beyond the bodies, and that all perceptible body is in place. One might even come to believe that Hesiod was correct to make Chaos, or the yawning chasm, the first of all things that came into being. Hesiod began his theogony in this way on the assumption that there must be a space or room for the beings before anything else can exist, an assumption rooted in the belief, which he shared with the many, that all things are in a place. Aristotle adds that if Hesiod is right, the power of place would be something marvelous; for if nothing else can exist without place, while place in turn does not perish with the destruction of the things in it, it must indeed be the first of all things. Yet this magnification of the status of place would seem to be only an extreme consequence of the common view according to which place is a universal container for all moveable bodies, a container that itself remains unaffected despite the changes among the bodies it contains.[5]

After presenting these arguments that place exists, Aristotle then turns to the perplexities regarding what it is, perplexities, to repeat, that compel us even to wonder whether there is such a thing. With one exception, these perplexities are not arguments of his predecessors, to which he might feel he must respond, but rather difficulties that he raises himself because of their intrinsic importance. He begins by asking whether place is a kind of bulk or volume of a body or else some other kind of nature. He asserts that it has the three dimensions of length, breadth, and depth, by which all body is defined, so that it would seem to be something bodily; and yet he says that it cannot be a body, for in that case two bodies, the place and the body it contains, would coincide. Aristotle's second perplexity brings out the difficulty of distinguishing a place, or the space that a body occupies, from that body itself. He argues that if there is a place of a body, then so too must there be a place of a surface, a line, and a point. But a point, he continues, does not differ from the place of a point, and so neither should the place of any of these other things, including the body, be anything apart from the things themselves. Aristotle calls the existence of place into further question by arguing that as something bodiless that nevertheless has magnitude, it can be neither an element, whether bodily or intelligible, nor composed of the elements. For the elements of the perceptible beings, he says, are (also) bodies,[6] while no magnitude comes into being from the intelligibles. His next claim, which calls into question the earlier argument that place is somehow responsible for the natural motions of bodies, is that none of the four kinds of causality belongs to it: it is, he says, neither the material of the beings, nor their form, nor their end or goal, nor an initiating source of their motions. Aristotle then brings up Zeno's argument that if place is one of the things that are, and if everything that is, is in a place, then there would have to be a place of the place, and so on to infinity, which seems absurd. And finally, he raises a difficulty based on the common assumption that the place of each body is neither smaller than nor greater than the body itself. For this equality of place with the body it contains seems to entail that the place of growing bodies must grow along with them, and this too seems absurd.[7]

We see that most of these arguments have fastened on the difficulties in the common view of place as a spatial extension, or a kind of room, which is occupied by the body that coincides with it at any given time.[8] But Aristotle has also argued that place is not the cause of anything, and in the course of this argument, he claimed, among other things, that it is neither matter nor form. He made these claims without discussion, as if they were evident truths; but in the immediate sequel he returns to these suggestions that a body's place is its matter or else its form, and he now brings forward a number of arguments to show that it can not be either of them. He gives all these arguments, moreover, despite prefacing them with the remark that it is easy (οὐ χαλεπὸν, 209b22) to see that place is neither matter nor form. And what is more, even after this thorough refutation of the two suggestions, he returns to them again, and again rejects them, as part of the argument that leads up to his own proposal of a definition of place. The fact that Aristotle considers, and reconsiders, the notions of place as matter or else as form, in the face of his own assertion that it is easy to see that it is neither of them, is perhaps the most surprising feature of this whole discussion. Let us try to see, then, why he dwells at such length on these particular misconceptions.

Aristotle continues his account of the perplexities regarding place by distinguishing the common place, which contains all bodies, from the private or particular place in which each body is primarily located. Thus you, he says, are in the world (ἐν τῷ οὐρανῷ, 209a33) because you are in the air, which is in the world, and you are in the air because you are in (or on) the earth, and you are on the earth because you are in the place that surrounds nothing more than you. Now Aristotle seems to suggest that only this last, or utterly particular, place is the place of a body in itself, and that it is only by virtue of being in such a place primarily that it can also be said to be in a larger one. This assumption, at all events, is the basis of most of his subsequent discussion; and it leads, in particular, to the suggestion that if place is what first surrounds each body, it would be a kind of limit, so that the place of each thing could be thought to be its form or its figure (τὸ εἶδος καὶ ἡ μορφὴ ἑκάστου, 209b3), by which the material of its magnitude is delimited. On the other hand, he continues, insofar as place is thought to be the extension of a magnitude, it could be thought to be its matter. For by the extension of a magnitude Aristotle means, in this context, an indefinite extension, one that is surrounded and made definite by form, as by a limiting surface. And such, he says, is matter or the indeterminate, on the grounds that nothing else is left—in the sphere, for instance—when its limit and its characteristics (τὸ πέρας καὶ τὰ πάθη, 209b10) are removed.[9] Aristotle adds that it is with these considerations in mind that Plato spoke in the *Timaeus* of matter, or of that which participates in the intelligible, as being the same as space or place.[10]

While the authority of Plato seems to give a certain weight to the identification of place with matter, it is harder to see any merit in the suggestion that it is a body's form.[11] But the need to be clear about what matter is in order to know whether or not it is the same as place, together with the intimate connection between matter and form, may be sufficient reason for considering both of these two suggestions. As Aristotle says, it would reasonably be thought difficult to know what place is if it is either matter or form, since these require the highest contemplation, or the keenest scrutiny (τὴν ἀκροτάτην ἔχει θέαν, 209b20), and it is hard in particular to know either of them without the other. Aristotle adds, however, as I have already mentioned, that it is easy to see that place cannot be either matter or form. For neither of these, he says, is (ever) separated from that thing whose form or matter it is, whereas its place can be, as for instance when air and water exchange places with one another. Such displacements among bodies, he continues, show that the place of each thing is neither a part of it nor a condition that characterizes it (οὔτε μόριον οὔθ' ἕξις, 209b27), but instead is separable from it. Aristotle confirms this rejection of the two suggestions by adding that place is thought to be something like a vessel, which he calls a moveable place, and which also, he says, is nothing of, or belonging to, the thing.[12]

Aristotle's argument has thus distinguished the place of each thing, as something separable, from its matter and its form, which are understood to be inseparable from it—apparently on the grounds that they each belong to the thing, either as a part or else as a condition that characterizes it. Yet though it is clear that a condition of a thing, as for instance the health of a body, cannot be separated from the thing itself (except in thought), this is not so obvious in the case of the parts. For a composite whole can sometimes be broken down into its parts without any damage to the parts themselves. And to return to our original question about matter and form, though it seems reasonably clear that the form of a thing cannot be separated from it—at least not if we mean by "form" the kind of contour that could conceivably be identified with place—there are doubts that arise with regard to matter. For though matter, according to Aristotle, is necessarily bound up with form, it might still be separable from any one particular form or thing.[13] Indeed, if air and water share a common material, as Aristotle himself sometimes suggests that they do, this matter would retain its identity in the transformation from the one element to the other, and it would thus be separated from the first one.[14] Now even if this were true, it would not show, of course, that matter is place. But the possibility does seem to show a weakness in Aristotle's argument as to why it is not.

This difficulty helps to explain why Aristotle proceeds to introduce a number of other arguments, beginning in particular with a partial repetition of the first one, in which the separability of a thing's place from the thing itself is used to distinguish it only from the form of the thing, and no longer from its matter. The distinction between place and matter is now said to be instead that a place surrounds or encompasses the thing whose place it is, whereas its matter does not. Now as Aristotle's analogy between a place and a vessel has already indicated, he means that the place of a thing encompasses it, not in the sense that a circumference does a circle, but rather as something distinct from the thing itself. Accordingly, he adds that what is somewhere is always thought to be something that has something else external to it. And the thought that the place of a thing, as opposed to its matter or its form, is something external to it seems also to be the primary basis for Aristotle's next two arguments. Thus, he asks how anything could move upward or downward to its own, or proper, place if place were either its form or its matter. And he argues that if a thing's place were something in it, as its form or figure (μορφή, 210a6) and its matter both are, then there would have to be another place, that is, a place in some other sense of the word, for this one; for whenever a thing moves from one place to another, its matter and its form also move, or change place, along with it.[15]

Aristotle has now repeated the claim that the form of a thing is inseparable from it, though he no longer says this about its matter, and he has also added a number of arguments that do not require the assumption that either of these belongs to the thing inseparably. But in the next and final argument of this series, he apparently returns to the claim of inseparability, with regard to matter as well as form. He begins this argument by asserting that when air is changed into water, the original place no longer exists. But what sort of perishing, he goes on to ask, can there have been? Now it is best, I think, to interpret this somewhat difficult argument in the light of its context, in which Aristotle has been arguing that place is neither matter nor form. Accordingly, its major premise, that the transformation of air into water entails the destruction of its place, is based on a view of place as matter or else as form. Aristotle would again be implying, then, that both the matter and the form of a thing are inseparable from it, and he uses the commonsense assumption that place is not destroyed in the transformation of elements in order to reject these two hypotheses about what it is. Now whether or not I have correctly interpreted this argument, Aristotle's commitment to the inseparability of matter, in particular, is confirmed later, when he returns to the hypothesis that place is matter. For in again rejecting this suggestion, he merely asserts—with a reminder that he has said this before—that a thing's matter is neither separable from the thing nor does it surround it.[16] But if Aristotle is so confident that matter is inseparable from the particular thing that it belongs to, why did he seem to retreat from this claim earlier? And how would he respond to the counterclaim that is based in part, at least, on statements of his own?

Aristotle provides no explicit answer to this question about the basis for his assertion of the inseparability of matter. Instead, he goes on to supplement his account of the perplexities regarding place[17] with a summary of the various ways in which one thing is said to be in another, along with a lengthy discussion of whether anything can be in itself. He uses his account of the different senses of the word *in* in order to respond to Zeno's objection that place itself must be in a place. But the conclusion of this whole section suggests that it is also meant as a continuation of the preceding discussion, for it reasserts that place is neither matter nor form. And yet this conclusion is somewhat puzzling, since nothing in this new section has added any weight, or at least not evidently so, to the grounds for rejection of either of these hypotheses.[18] Let me propose, then, that this section might also have an ulterior purpose, namely, to strengthen the case against the view of place as matter by sketching—between the lines, as it were—the outline of an argument that the matter of a thing is inseparable from it.

To see that this is so, let us turn first to Aristotle's discussion of the question whether something can be in itself. He says that this is not possible, except in the sense that one part of a thing can be in another and that we sometimes speak, in such a case, of the whole thing being in itself. Thus, for instance, when a certain measure of wine (an amphora) is in a certain kind of vessel (an amphora), the amphora of wine can be said to be in itself. Now this manner of speaking presupposes that we can apply a term (such as *amphora*) to a composite whole although our primary reference is to one or another of its parts, and to help illustrate that this is done, Aristotle offers a pair of additional examples. He says that someone can be spoken of as white, although it is the surface of his body that is white primarily, and likewise that he can be spoken of as a knower because of the knowledge in the rational part of his soul. These two examples, moreover, turn out to be particularly relevant in connection with our unresolved question about matter, since they call attention to body and soul as such, which Aristotle regards as being related to one another as matter and form.[19] Here, it is true, he speaks less precisely of the soul and the body as parts of a human being, and he even suggests an analogy between the soul in the body and the wine in the amphora.[20] And yet this very analogy leads us to raise the question of whether the human soul and the human body can exist apart from one another, as the wine and the amphora clearly can. And whatever we may think regarding the fate of our souls, we know that the human body can not exist separately, but rather begins to decompose once the human being has died. Thus we have found one case, at least, in which the matter of a being is clearly not separable from the being as a whole. More generally, since Aristotle speaks of the matter of a being as that in which form is immediately present, as distinguished from any underlying constituents into which this matter can be broken down,[21] the matter of all living beings, at least, is inseparable from those beings. Still, the question remains of whether what is true in these cases is universally true in nature, and in particular, whether it is true in the case of the elements—such as water and air, for instance, which serve so frequently as illustrations in this discussion of place. Can Aristotle's claim that the matter of each thing is inseparable from it be justified even in these cases? Or must we say, as has been suggested earlier, that the elements share a single, common matter that retains its identity throughout their transformations into one another?

To answer this question, it is helpful to note that in the present context, in his listing of the various ways in which one thing can be in another, Aristotle uses the term *form* in several senses: he speaks of it as something, such as health, which is present in matter; but also as a species, or the class to which a thing belongs; and as a definition, or what the definition of a thing intends.[22] And these latter two senses are in fact the more precise ones, since a form, by its being the defining aspect of a thing, is distinguished from any mere condition such as health. Thus, for instance, soul, or the form of a living being, is not a mere condition in which its body may be found, but is instead the very being, in a sense, of such a body.[23] And so even if it were not the case that corpses decompose, what survived the being's death would no longer be the same body. It could be called the same body only in an equivocal sense, since it would no longer have the powers that had characterized it as the being it was. For this reason alone, then, the body, like the soul, is an inseparable aspect of a living being. And we can also say in general that both the matter and the form of any being are inseparable from it, since apart from that being they would no longer be what they are. Now in the case of the four elements, it is admittedly hard to distinguish formal and material aspects. But the analysis I have outlined suggests that matter, as well as form, must be inseparable from the element to which it belongs. And in fact, Aristotle argues explicitly at the end of *On the Heaven* that since any amount of each element has the same upward or downward tendency in relation to each other element, they must therefore have four different kinds of matters.[24] Thus, even in the case of the elements, matter is inseparable from the being to which it belongs, since like form, it becomes manifest only as an aspect of that being. And by understanding the matter of the elements in this way, we no longer have to try to imagine it as some characterless substrate that maintains its "identity," whatever that would mean, throughout their transformations into one another.[25]

The view of matter that I have outlined here justifies Aristotle's claim that the matter of each thing is inseparable from it; and we have thus completed our interpretation of his argument that place is neither matter nor form. But despite the importance of a discussion of form and matter, the question remains, given the ease of distinguishing either of these from place, of why this discussion is given such a prominent role in an account of place in particular. What of importance does it contribute to our understanding of place? Let us continue to keep this question in mind as we turn from Aristotle's treatment of the perplexities regarding place to his attempt to say positively what it is. This attempt, as we noted earlier, begins from the assumption that place does indeed have the attributes that are truly believed to belong to it in itself. We are asked to assume, in particular, that place surrounds or contains the thing whose place it is and that it is not anything that belongs to the thing. Next, we are asked to assume that the primary place of a thing is neither smaller than nor greater than it is, but rather, as Aristotle says elsewhere, that it is equal to it. We are also asked to assume, regarding this primary place of each thing, that it can be left behind by it, while remaining as something separate. And finally, we are asked to assume that all place includes the up and the down, and that every body moves naturally either upward or downward to its own or proper place, where it then naturally remains.[26]

After listing these preliminary assumptions about place, Aristotle then presents the outline, which I quoted at the beginning of this chapter (cf. pp. 77–78), of that "most beautiful" manner in which one might hope to explain it. And he begins his attempt to carry out this program with the observation that there would be no inquiry regarding place if there were no locomotion, that is, change of place. In the absence of such motion, he implies, we might speak—if speech were possible—of various parts in a larger whole, but we would never distinguish one thing from another so completely as to say that anything was in a place. Thus, he continues, we suppose that the heaven (τὸν οὐρανὸν, 211a13–14), more than anything else, is in a place, since it is always in motion. Aristotle goes on to say that what is in motion is so either in itself, and actually, or else only by concomitance. And of things that are in motion by concomitance, there are some, he says, such as the parts of a body or the nail in a ship, that have the potency to move in themselves, and others, such as whiteness or science, that do not. For these (conditions), he explains, change place (only) in the sense that that in which they are (ultimately) present does so.[27]

Aristotle's emphasis on the connection between place and bodies that change their place is in keeping with our assumption that the primary place of each thing can be left behind by it. He next turns his attention to the additional assumption that this primary place is also equal to the thing in it, while still sharing with place in the broader sense the characteristic of surrounding that thing. We say, he tells us, that something is in the world (ἐν τῷ οὐρανῷ, 211a24) as a place because it is in the air, which is in the world; and we say that it is in the air, though it is not in all the air, because of the extremity of air that (immediately) surrounds the thing in question. For if all the air were its place, he continues, the place of each thing would not be equal to it, as it is believed to be. Aristotle adds, however, that when a container is continuous with what it contains, this latter is not said to be in the former as in a place, but is spoken of as a part in a whole. On the other hand, he says, when the container is divided from and contiguous with what it contains, this latter body is primarily in the extremity of the container. He contends that this extremity is neither a part of what is in it nor greater than its extension, but, rather, equal to it, on the grounds that the extremities of contiguous bodies are together (literally, "in the same," ἐν . . . τῷ αὐτῷ, 211a33–34; cf. 226b21–23). And finally, to help distinguish between these two kinds of containment, he goes on to say that what is continuous with its container, as the hand is with the body, does not move in it, but rather with it, whereas that which is divided from its container, as for instance water in a jar, does move in it and does so equally whether or not the container is in motion itself.[28]

Though Aristotle has surely suggested in this passage that the extremity of a surrounding body, at which it is in contact with the one it surrounds, is the place of this latter one, he has not yet said so explicitly. Moreover, he does not say this in the immediate sequel, but rather treats this suggestion as only one of four alternatives, including even the previously rejected ones of matter and form, which he now mentions as the possibilities for what place might be. And it is merely on the basis of having rejected the three other alternatives that he finally concludes that it must be this remaining one.[29] Now it is puzzling that Aristotle should use this argument by exclusion as the grounds for his proposal of a definition. He does, it is true, offer a kind of formal justification for it, since he begins with the claim that place must necessarily be one of a number of alternatives; but he does not explain how he came up with his own list of these, and he clearly suggests that it may not be exhaustive (σχεδὸν . . . τέτταρά ἐστιν ὧν ἀνάγκη τὸν τόπον ἕν τι εἶναι, 211b6–7, emphasis mine).[30] He does not even explain, moreover, the basis for his disregard of the alternative he had mentioned previously that there might not even be such a thing as place. And so let me suggest that he deliberately employs a visibly inadequate argument for his proposed definition, and that he does so in order to call attention to some difficulties in the proposal itself. For even if it avoids circularity, that is, if the togetherness of the extremities of the two bodies is not in the strict sense sameness of place,[31] the fact remains that the extremity of the surrounding body is not equal, as Aristotle has claimed it is, to the extension of the body that it surrounds. For the extremity of a body, or a two-dimensional surface, is not even comparable in terms of size to a whole body, which is three-dimensional. Thus, a definition of place as such an extremity does not truly correspond to our assumption that the place of a body is equal to it.[32] Secondly, there is a difficulty, at least whenever the surrounding body is not a solid, regarding the premise that place is separable. For even though the surface of air, for instance, in contact with a moveable body can be left behind by that body, it does not survive, as a distinct surface, after the body has moved away. To the extent, then, that such fluid surfaces are meant to be covered by the proposed definition of place, it fails to correspond to another one of our preliminary assumptions. And finally, Aristotle has explicitly allowed that the container in whose extremity a body is primarily found might itself be in motion. But if this extremity is the place of the contained body, then it would seem to follow that place could be in motion, even if only by concomitance, and this seems odd. For we tend to think of place as something unchanging.[33] That these are genuine difficulties with the proposed definition is confirmed, I think, by the character of the one new alternative that Aristotle first considers here before explicitly making his proposal. For this new alternative at least claims to make of place something equal to the body it contains, as well as being both permanent and unmoved. What it says is that a place is an interval or extension between the extremities around a body, an interval that remains forever and into which various bodies can enter in turn.[34] Now this view of place, though it is here stated explicitly for the first time, is of course not really new to Aristotle's discussion. It was already suggested by the initial arguments that place exists, and many of the subsequent perplexities have presupposed that something more or less like this is what people mean by the term. It is not surprising, moreover, that this view of place should be widespread, since it is indeed plausible or seductive. And given the difficulties that we have pointed to in Aristotle's own account, it at least makes sense that he should delay his proposal until after he has provided a critique of this alternative.

Aristotle says that there are, roughly speaking, four alternatives of which place must necessarily be one. It is, he says, either form (μορφή, 211b7) or matter or a certain interval—namely, the one between the extremities—or else it is the extremities, if there is no such interval apart from the magnitude of the body that comes within them. And it is manifest, he continues, that place can not be any of the first three of these. Now Aristotle prefaces his arguments against these views of place with a brief account of why they might seem persuasive. Thus, he begins by saying that it is because it (also) surrounds that form (ἡ μορφή, 211b11) is believed to be place. For the extremity of the surrounding body, he continues, is coincident (literally, "in the same," ἐν ταὐτῷ, 211b11; cf. a33–34) with that of the one it surrounds; and as a consequence, they might appear to be the same thing. But Aristotle replies that these extremities differ by their not belonging to the same thing, since the form is the limit of the thing in place, whereas the place is that of the surrounding body.[35]

Having thus again dismissed the suggestion that place is form, Aristotle now turns to the one that speaks of it as the interval between the extremities. He says that because the surrounded body often changes (its place) while the one that surrounds it remains—as for instance when water flows from a vessel—what is in between is believed to be something, an interval, on the supposition that this is something apart from the moving body.[36] Now it is not at all clear why such an occurrence should lead people to believe that there is some interval within the vessel apart from the body or bodies that it contains. But for now let us pass over this difficulty, since Aristotle will offer much the same account of this belief a bit later, and his additional explanation there, in conjunction with the intervening argument, makes it easier to understand his assertion.

In response to this view of place as an independently existing interval, Aristotle begins by denying that there is any such thing between the extremities of a surrounding body. Rather, he says, some chance body, among those that move and whose nature it is to be in contact (e.g., with the surrounding body), falls within it.

He continues by arguing that if there were some interval that existed by nature and that remained (permanently)—as those who believe in these independent intervals suppose that they do—there would have to be infinitely many places within the same one. For when the water and the air move (μεθισταμένου, 211b21), he says, (as for instance when air displaces the water in a vessel, or when the vessel is merely carried to a new place,) all the parts within the whole (of the moving body or bodies) will do just what all the water in the vessel does, that is, they will change place.[37] And though Aristotle does not say so explicitly, it would seem that these infinitely many places of the infinitely many parts would have to be an actual infinity—the kind that he argues elsewhere does not exist[38]—even though the parts themselves would not. For the claim that the interval within the vessel is independent of it, and thus actually in existence, whether or not the vessel remains there to surround it, implies that the places of the parts of the moving bodies are also actually in existence, even though these parts themselves exist only potentially until they are somehow marked off from the whole.[39]

Aristotle continues his rebuttal by asserting that the view in question would make of place itself something that moves, or changes its place, so that there would be another place of the first one and many places would be together. The situation he has in mind here is that of a vessel with fluid in it being carried as a whole to another place; when this happens, he argues, the interval within the vessel—or the place of the fluid, as it is claimed—would also move to another place, and thus the two places would coincide. Now the advocates of place as an independent interval might indeed refuse to acknowledge that such a place is ever moved along with a body.[40] But though there is no internal inconsistency in this refusal, it does not correspond to our experience of the world. For it would compel us to deny, for instance, that equipment stored in a moving vehicle could ever remain in the same place, and this does not make sense.[41] Or as Aristotle goes on to say, "when the whole vessel moves, the place of the part, in which it moves (μεθίστηται, 211b26), is not different [from moment to moment] but the same. For the air and the water or the parts of the water move (or, 'displace one another,' ἀντιμεθίσταται, 211b27; cf. 209b25) in the place in which they are, and not in the place in which they come to be, which is a part of the place that is the place of the whole world."[42]

Now though Aristotle has here criticized a view of place that would make of it something that moves, he acknowledges himself that this is true of it, at least in a sense. For even on his own account, the place that remains the same when the whole vessel moves must, if only by concomitance, be in motion as well. Still, his main argument here against those who think of place as an interval is that on their view the moveable place must be in (i.e., coincident with) another place (cf. ὥστ', 211b24; 216a26–b12), and this he is not compelled to admit. For if a place is the adjacent surface of a surrounding body, it is not, unlike the body to which it belongs, surrounded by another body or in a place.[43] But if, on the other hand, it is an independently existing interval—and thus ultimately, as Aristotle has shown, an actual infinity of such intervals—any moveable place would presumably coincide with a place of its own just as much as would a body or a part of one.[44]

There is, however, at least one further difficulty in Aristotle's response to this view of place as an interval. For in his account of the moving vessel he has not said precisely that its contents remain in the same place, but rather that they "move" in it, and yet their motion would seem to be a change of place. Now if their motion were exclusively motion by concomitance, as the whole vessel is moved from place to place, this statement would not create so much of a problem, or at any rate not a new one. But Aristotle suggests that he also has in mind cases in which the contents of the moving vessel move on their own, whether through the displacement of water by air or else though reciprocal displacement among the parts of water (see, again, ἀντιμεθίστοται, 211b27). And yet if this is so, how can these contents be said to move in the same place, at least in the primary or truest sense of the term "place"?

We will return to this difficulty shortly. But for now, let us continue with Aristotle's consideration of the alternative views of place. The third and last alternative that he rejects here is of place as matter. He says that matter might be thought to be place if one were to consider (the transformations) in something continuous and at rest. "For just as [we imagine that] if there is alteration, there is something that is now white that was previously black and now hard that was previously soft (for which reason we say that matter is something), so also place is believed to exist because of some similar imagination [διὰ τοιαύτης τινὸς . . . φαντασίας]; except that the former [is believed to exist] because [we imagine that] what was air is now water, whereas place [is believed to exist] because [we imagine that] where there was air there is now water. But matter, as was said before, is neither separable from the thing nor does it surround it, whereas place has both [of these attributes]."[45] Now in reasserting that matter is inseparable from the particular thing it belongs to, Aristotle confirms that he does not believe that there is a common matter of the four elements. But what is new here is his explicit suggestion that the notion of such a permanent substrate is an imaginary one. We imagine, he suggests, that there must be a substrate that remains unchanged in every alteration, including even those, such as the transformation of air into water, where no perceptible body remains the same. And though Aristotle does not say so explicitly, he also invites the thought, it seems to me, that our belief in this imagined substrate stems from a *wish* to believe in it, to believe that at least the simplest of the perceptible bodies that constitute our world are rooted directly in a permanent and changeless substrate. If this is Aristotle's thought, moreover, the parallel that he draws between the imagination that leads to the belief in matter, in this sense, and the one that leads to the belief in place would suggest that this same wish to believe in something stable is also at the root of the belief in place, or of the belief, at any rate, that it is entirely free from change.[46] And the importance of this suggestion would allow us finally to understand why he has given such a prominent role in his account of place to a discussion of form and, more particularly, of matter. Now to be sure, this suggestion of mine about a wish to believe in place is only speculative. But in the sequel to this discussion of matter, Aristotle points to the need for at least some such speculation, as we shall see in our examination of the text.

Aristotle now concludes, on the basis of having rejected the first three of his alternatives, that place must be the remaining one, or the limit of the surrounding body, at which it is in contact with the moveable body that is in place. Although we have seen that place, on this view, does not fully correspond to all our initial assumptions about it, Aristotle's refutation of the alternative notion of an independently existing interval suggests that he has at least put his finger on something that corresponds to them more nearly than anything else that really exists. And to the implicit objection that this view of place as the mere limit of a surrounding body does not give to it the importance it seems to have, Aristotle goes on to suggest that we are deceived in believing that place is something great and hard to grasp (δοκεῖ δὲ μέγα τι εἶναι καὶ χαλεπὸν ληφθῆναι ὁ τόπος, 212a7–8; cf. 211a7–11). He says that this belief about place arises in part because matter and form appear along with it[47] and also because the change of the body in locomotion occurs within a surrounding body at rest. "For it appears possible that there is some interval in between other than the moving magnitudes."[48] Now we recall that Aristotle has already suggested that the fixity of the body surrounding the one in motion is what leads to the belief in an independently existing interval, and in discussing that suggestion I said that it did not make evident sense. But now, I think, we are in a better position to understand what Aristotle has in mind. The rest or fixity of the body that surrounds the one in motion does not, in fact, give rise by itself to the belief in an interval apart from bodies. Indeed, Aristotle implies as much in the immediate sequel, for he adds that the apparently bodiless character of the air in an "empty" vessel is also a contributory cause of this illusion. But even these two causes taken together do not suffice to explain the prevalence of the belief in place as an independent interval. And by the inadequacy of his explicit account of the origin of that belief, Aristotle invites and even compels us to supply some additional cause on our own. Now this other cause, in my view, is the wish that I referred to earlier, the wish to believe in something independent and wholly unchanging as the container of our world. The wish to believe in such a permanent container is the crucial factor, I contend, in engendering the belief in place as an independent interval. And the fact that by holding this belief about place we also come to think of it as "great" helps to confirm, I think, the importance of this wish for stability, for the stability it wants to believe in is of something great. And what the fixity of the surrounding body contributes to this belief is only the manifest and partial stability that helps make it possible for us to imagine, and thus actually to believe in, such a completely stable interval behind the scenes.

Throughout his discussion Aristotle has frequently used the example of a vessel, or its inner surface, to illustrate what he means by place. He has called a vessel a moveable place, and he has at least suggested that these moveable places are a subclass of place in the broader sense.[49] But now, however, he takes the argument in a new direction by insisting that place, as opposed to a vessel, must be immoveable. He says that just as a vessel is a moveable place, so place is an immoveable vessel. And accordingly, he continues, when something moves or changes in a surrounding body that is in motion—as a boat, for instance, moves in (the flowing water of) a river—it uses this body as a vessel, rather than as a place. Aristotle goes on to claim that place wishes to be immoveable (βούλεται δ' ἀκίνητος εἶναι, 212a18), and that therefore the whole river is rather the place (of the boat), since the whole is immoveable. And on the basis of these claims, he proposes a revised definition of place as "the first, immoveable limit of the surrounding [body]," that is, the immoveable limit at which it is in contact with the body it surrounds.[50]

Now the demand that place be immoveable, though it is presented here explicitly for the first time, is not so surprising in itself, since we do tend to think of place as a stable background for locomotion. Still, my account of the argument that has led up to this demand makes us wonder whether it can be met. And Aristotle's revised definition of place, as the immoveable inner surface of the surrounding body, presents this difficulty in an especially acute form, coming directly as it does after an example in which he has more or less told us that there is no such surface.[51] Since the water that surrounds a moving boat in a river is continuously in motion (so that the surface at which it touches the boat is also in motion by concomitance), Aristotle has suggested that the place of the boat is not this changing surface, but rather the immoveable river as a whole. At least in this case, then, he has preserved the view that place is immoveable only by abandoning the requirement that each place must be the place of only one particular body. And even the requirement of immovability is met only in a sense, since the river that remains unmoved as a whole does so despite the fact that each portion of its water is continuously in motion and is being replaced.[52] Moreover, this example of a boat in a river is by no means an unusual one, for the air that surrounds us is also in constant motion. And even though we can stand still, so that the inner surface of this air remains unmoved—in a sense roughly similar to that in which a whole river, despite its flowing, remains unmoved—we have already noted that this surface ceases to exist as an actual surface as soon as we walk away.[53] More generally, if we disregard for the moment the special cases in which moving bodies do not change places as wholes, at least part of the surface at which a surrounding body or bodies are in contact with the moveable body must be moveable or changeable itself.[54] And to this extent, at any rate, Aristotle's definition of place as the first, immoveable limit of the surrounding body looks more like a statement of what we might wish for it to be than a definition of anything real. For we do not merely wish that there be an independent and wholly unchanging container of our world; we also want this comprehensive place to include equally unchanging parts that fit exactly to each body, including our own, and in which we can be at rest.[55] But Aristotle suggests that this wish for stability is unattainable. And from this perspective, I think, we can understand why he says here that place "wishes to be immoveable," rather than that it is immoveable in fact.

Even once we accept, however, that place in general is not wholly unchanging, it is still not clear in what sense we are to understand Aristotle's claim that the whole river is the place of a moving boat. For since, as we have noted, this place contains more than the one boat, it would seem not to be a particular place in the sense that Aristotle has laid out, but rather a common one, in which a body is found only by virtue of being primarily in one of the other kind. If, however, we look back to the passage in which he first distinguished these two senses of place, we note that the only thing that he spoke of as the common place was the place of all the bodies (in the world), and that he did not say in general—as opposed to suggesting it through his example—that the particular, or primary, place of a body must fit it exactly.[56] Thus, his claim here that the whole river is the place of the moving boat need not imply that there is another, more primary place—whether moveable or not—in which it is found.[57] And it makes sense that there is none. For since a moving body is not, according to Aristotle, ever actually at any midpoint along its path, neither is it actually at any place during its motion other than the whole place in which it moves.[58] And from this point of view, we can perhaps also answer our earlier question as to why Aristotle had suggested that the contents of a (moving) vessel displace one another in a single place.[59] For if we consider, for instance, the portions of water that move randomly and splash about, it makes little sense to say that they remain "in" a surface with which they are only occasionally in contact. But it does make sense to say that they displace one another while remaining within the vessel as a whole.[60] And more generally, the notion of place as the inner surface of a surrounding body gives a sense of exactness that is at least in many cases untrue to the phenomena. It is often better to think of a place as a bodily being, and even a moveable one, considered as a whole.[61]

Although the definition of place as the first, immoveable limit of the surrounding body is not a correct statement of what place generally is, and although it fails even in principle to correspond to our assumption that a place is equal to the thing in place, this is not to say that there is no truth to it. This definition may still give the most satisfactory possible sense of the term *place*, a sense that is more or less fully realized in some actual places, however uncommon these might be. Thus, for instance, the unchanging surface of a depression in the earth would be the place of the air or of the rainwater that this earth surrounds, even though it does not surround them completely. And if we ask about the places of the four elements considered as wholes, it would seem—since these wholes do not move from their places—that they are each completely surrounded by a body whose inner surface, as a whole, remains unmoved. Accordingly, Aristotle turns his attention to the places of some of these elements. He says that his definition of place explains why the middle of the world and the extremity toward us of the circular locomotion (i.e., of the revolving heavenly sphere) are believed by everyone, more than anything else is, to be the up and the down in the chief sense: for the one, he says, remains forever, and the extremity of the circularly (moving body) remains situated in the same way. Now of course whatever is meant by "the middle of the world," this would not seem to be a limiting surface; and perhaps partly for this reason, Aristotle goes on to restate in his own name what is meant by "up" and "down." Since, he says, the light and the heavy are what move up and down, respectively, by nature, both the limit that surrounds toward the middle and the middle itself are down, and both that (which is) toward the extremity and the extremity itself are up. He may mean by this difficult statement that the surface of whatever body surrounds the heavy one at the middle of the world is the place that is down, and that "the middle itself"—that is, the earth (since no center point exists in actuality)—is also spoken of as being down; and since by contrast to the midpoint of the world, the extremity (toward us of the heavenly sphere) is a surface, what he speaks of as "the extremity itself" may be this place of the lightest body, and that which is toward the extremity may be the body (i.e., the element fire) in this place.[62] On this view, the places of the earth as a whole and of the sphere of fire as a whole are the inner surfaces of the immediately surrounding bodies, which surfaces remain always unmoved or at least situated in the same way. Accordingly, Aristotle goes on to say that for this reason place is believed to be a kind of surface and, as it were, a vessel and a container. And he adds that place is also together with the thing (in place), since the limits are together with the limited.[63]

My interpretation of Aristotle's statement about what is down and what is up has denied the apparent parallelism between the expressions "the middle itself" and "the extremity itself," since I took the former to refer to the lowest body and the latter to the highest place. If instead we try to preserve this parallelism, as well as the parallelism between "the limit that surrounds toward the center" and "that [which is] toward the extremity," the latter of these two expressions would also refer to a limiting surface and would designate the inner surface of the heavenly sphere as the highest place; but then "the extremity itself," if it is the body in that place, would have to be on this side of what is "toward the extremity" and that is difficult.[64] Perhaps, however, "the extremity itself" refers not to the body in the highest place, or to the fire within the concave surface of the lunar sphere, but rather to the body above it, that is, the heavenly body.[65] This suggestion has the merit, at any rate, of calling attention to this heavenly body and to the difficult question of where it is, on the assumption that place is an extremity of a surrounding body. And this question will come to the fore in the following section of Aristotle's discussion, together with the related question regarding the place of the whole world.

Aristotle continues by saying that if a body has some external body surrounding it, it is in a place, and that if it does not, it is not. From this it follows, among other things, that the world as a whole has no place, and to drive home the truth of this paradoxical conclusion, Aristotle extends it to the imaginary case of a world wholly composed of water, that is, of something that seems to require a containing body. Even if water, he says, should come to be such (i.e., not surrounded by an external body), its parts will move, since they are surrounded by one another; but the all will move in one sense, but in another sense not. "For as a whole, it does not change its place all together, but it moves in a circle—for there is this place of its parts—and some [parts] do not move up and down, but in a circle, whereas others, those that admit of condensation and rarefaction, move both up and down."[66] Now in this last statement Aristotle seems to be speaking again of the real world, rather than of an imagined watery one, and his claim is that it does not change place as a whole, even though the heavenly spheres move circularly and sublunar bodies move up and down. In saying, moreover, that the world does not change place as a whole, he does not mean that it remains at rest in the same place, but presumably that it is neither in motion nor at rest, and that it thus has no place either to preserve or to change.[67] However, this last suggestion, though it is clearly in keeping with the overall purpose of Aristotle's argument, is called into question by some ambiguity as to whether he thinks that the world moves in a circle. For first he says that it does, but in his restatement he refers only to some of its parts as doing so. Now since the earth, at least, does not move in a circle (according to Aristotle), it would seem that this second formulation is the more precise one. But why, then, does he say in the first place that the all moves in a circle? The commentator Simplicius has suggested a possible answer to this question, namely, that Aristotle is here disregarding the earth, as well as the region immediately around it, on the grounds of its being too insignificant to be a genuine part of the whole.⁶⁸ Yet even if this suggestion is true, the problem remains of explaining how the world can move in a circle if there is no other body in relation to which it can move. Perhaps it is better, therefore, to interpret the claim that the all moves in a circle to mean merely that some of its parts do, as when we say that a man is injured because of a wound to his chest.⁶⁹ But even if it is only the heaven that moves in a circle, there is still the question of how it can do so, since according to the definition of place as a limit of a surrounding body, it would seem to have no place in which to move.⁷⁰ We recall Aristotle's earlier assertion that locomotion is what leads to the thought of place and that in particular we suppose that the heaven (τὸν οὐρανὸν, 211a13–14), since it is always in motion, is in a place more than anything else is.⁷¹ Can we now deny that the heaven even has a place, as his definition of place would seem to require us to do, without also denying that it moves? Let us keep this question in mind as we look at the continuation of the argument that the world has no place.

After distinguishing between being potentially in place and being so actually, and between things that in themselves are in place and those that are so only by concomitance, Aristotle now says explicitly that the world (ὁ δ' οὐρανός, 212b8) as a whole is not in any place, at least on the assumption that there is no body surrounding it. He means by this that it is not in itself in any place, for he goes on to mention both the world and the soul as examples of things that are in place by concomitance. The phrase "by concomitance" is apparently used here—in reference to the world, at any rate—as an equivalent to "by virtue of its parts."⁷² For its parts, Aristotle continues, are all in a sense in place, since one surrounds another on the circle. Now it is presumably the difficulty regarding the heavenly motion that explains why Aristotle limits himself to the claim that it is only "in a sense" that "all" the parts of the world are in a place. And the commentator Themistius has suggested that the outermost sphere of the heaven, the sphere of the fixed stars, is the one that is only in a sense in place, since it is not truly surrounded by any body, but only surrounded in the qualified sense of being in contact with the sphere of Saturn at its inner extremity.⁷³ And there is some merit to this suggestion. For in more or less the same sense that we speak of a vessel as surrounding the fluid in it, even though it does so from below, and only in part, we can also speak of a lower sphere as surrounding a higher one.⁷⁴ However, this suggestion fails to explain how the place of this outermost sphere will be unmoved, since the sphere of Saturn, like all the heavenly spheres, has (approximately) the same daily revolution as the fixed stars do. If the sphere of Saturn moves circularly in this way, how can its outer surface be the unmoved place for the revolution of the sphere of the fixed stars? Merely to say that this inner sphere remains as a whole in the same place is not sufficient, since this is equally true of the outermost sphere, whose rotational motion requires a place that does not rotate. Must we not say, rather, in order to make sense of the daily revolution of the heavenly body as a whole, that it is the sublunar region, and ultimately the earth or its surface, that provides the fixed place in or around which this motion occurs?⁷⁵

Now it is true that Aristotle never says all this explicitly. But as we shall see, his explicit discussion gives no satisfactory answer to the question regarding the place of the heavenly motion. And he has reason, moreover, to be reticent about his answer to it. For if my suggestion is correct, not only do the sublunar sphere and the heavenly sphere provide the places for one another, but the former of these places is the more important one, since only the earth or its surface has the ultimate fixity presupposed by all motion, including that of the "fixed" stars. And this is not to say that the earth is simply "beyond" the realm of the moveable, since every portion of it is moveable and even perishable.⁷⁶ Rather, we find that the attempt to interpret the experienced fixity of the earth, and the experienced motion of the heaven, in relation to some absolutely fixed place turns out to be futile. And thus precisely this *experienced* fixity of the earth becomes the ultimate perspective in relation to which motion and rest exist. Accordingly, when Aristotle says that the up and the down and the other differences of place exist not only in relation to us and arbitrarily (πρὸς ἡμᾶς καὶ θέσει, 205b33–34; cf. 208b12–22), but also in the whole itself, the whole that he has in mind is a whole that exists as such only in relation to our human experience. And yet his formulation points to this understanding of the whole without openly challenging the popular understanding or openly threatening the sense of security that it may give.⁷⁷

I said that Aristotle's explicit discussion gives no satisfactory answer to the question regarding the place of the heavenly motion. To see that this is so, and to see how he both conceals and calls attention to the difficulty, let me begin by quoting him.

> Therefore the upper [part of the world] moves in a circle, but the all is not anywhere. For what is somewhere is itself something, and there must also be something else beyond this, in which [it is and] which surrounds it. But beyond the all and whole there is nothing external to the all, and for this reason all things are in the world; for the world [ὁ . . . οὐρανὸς, 212b17] is perhaps the all. Yet the[ir] place is not the world [ὁ οὐρανός, 212b18], but something belonging to the world [τοῦ οὐρανοῦ τι, 212b18–19], its extremity, [which is] also in contact with the moveable body. And for this reason the earth is in the water, and this is in the air, and this is in the aether, and the aether is in the world [ἐν τῷ οὐρανῷ, 212b21], but the world [ὁ δ' οὐρανὸς, 212b22] is no longer in anything else.[78]

Now an obvious question regarding this passage is what Aristotle means by "the extremity" of the world, which he seems to say is the place of all things, and which he characterizes as being in contact with "the moveable body." It might appear from his language that what he has in mind is an immoveable outer sphere, whose inner surface would be primarily the place of the moveable sphere of the fixed stars.[79] On this reading, in the subsequent claim that the aether is in the world, the word *aether* would refer (as it popularly does) to the circularly moving heaven;[80] and Aristotle would be suggesting that its place, within the world, is the inner extremity of this immoveable body beyond it. Now this interpretation does, to be sure, provide an answer of sorts to the question of the place of the heavenly motion. Yet there is no evidence at all for the existence of an immoveable body beyond the visible and moving heaven, and Aristotle never says that he thinks there is one. Accordingly, an alternative interpretation of these last two sentences has generally prevailed.[81] On this interpretation, the word οὐρανός, which I have here translated as "world," and which almost certainly did mean "world" earlier in this argument, is translated instead as "heaven," as indeed it often must be; and its extremity in contact with the moveable body is interpreted as the lunar sphere, or its inner surface, which Aristotle had earlier said is believed by everyone to be the place above. In keeping with this suggestion, "the moveable body" in contact with the extremity is interpreted as the totality of the sublunar bodies, that is, those whose motion is rectilinear. And in the concluding sentence, the word *aether* is interpreted as having the Anaxagorean sense of "fire," so that Aristotle's assertion comes to mean that the element fire is in the heaven and that the heaven is not in anything else. Now although Aristotle elsewhere disapproves of Anaxagoras' use of the word *aether* to refer to fire,[82] this interpretation does have the advantage of not omitting a reference to this fourth of the sublunar elements, and it also of course has the great advantage of avoiding the rash hypothesis of an immoveable outer sphere. But it gains these advantages at the price of disregarding the whole question of the place of the heavenly motion, as if the only moveable bodies, or the only ones that needed to be in a place, were those of our sublunar region. And Aristotle's equivocation as to whether he is talking about the heaven or the whole world invites the reader, or at least those readers whose doubts are too easily put to rest, to pass over this difficulty. For his studied equivocation allows him to give the impression that the question of the place of the heavenly motion can be dismissed on the grounds that there is no place of the immoveable whole. And yet the reader who resists this temptation and who continues to ask the question will be led, I think, to answer it in the way I have proposed.

Having offered his definition of place, and having clearly spelled out the implication, at least, that there is no place of the whole world, Aristotle now asserts that all the perplexities regarding place can be resolved on the basis of this definition. We recall that he had listed these perplexities early in his discussion and that it was part of the original goal of his inquiry to resolve them. In now doing so, he apparently relies on the broader version of his definition of place, the one that defined it as a limit of the surrounding body but that did not yet insist on its being immoveable. For he begins by saying that there is no necessity for a place to grow along with the thing in place. And the implicit argument for this conclusion—namely, that the increase in the size of the surface of the surrounding body is not growth, but a mere concomitant of some other change in that body itself—is acceptable only if one does not have an immoveable surface in mind.[83] Aristotle continues with the assertion that there is no necessity for there to be a place of a point, since, as he allows us to infer, a point is not surrounded by any body. The next perplexity that he turns to is the one he had originally mentioned first, which had argued from the apparent three-dimensionality of place that it must be a body, but that it could not be a body, since there would then be two bodies in the same place. His response to this perplexity is a partial repetition of his earlier response to the view of place that it presupposes, or the view that place is a spatial interval. He says that there is no necessity for there to be two bodies in the same place nor indeed for there to be any bodily (i.e., three-dimensional) interval (between the extremities of the surrounding body), since what is between these extremities is some chance body, but not an interval for a body (σῶμα γὰρ τὸ μεταξὺ τοῦ τόπου τὸ τυχόν, ἀλλ' οὐ διάστημα σώματος, 212b26–27). And finally, he responds again to Zeno's objection that since a place is somewhere, it must be in another place, and so on to infinity.[84] Aristotle acknowledges that a place, since it exists, must be somewhere or in something. But this does not mean, he says, that it is in a place, but rather that, being a limit, it is in what is limited, i.e., in the body that surrounds the one in place. For not everything that is, he continues, is in a place, but only moveable body.[85] We note that despite Aristotle's claim here that all the perplexities regarding place can be resolved, he has failed to mention two of them, the one that argued that place could be neither an element nor composed of the elements and also the one that denied that it could be any of the four kinds of causes. His reason for not mentioning at least the former of these perplexities is presumably that he agrees with it, since it is a difficulty only for those who think that place is something more important than it is. For a surface of a body is not an element nor even a being composed of the elements, but merely an extremity of such a being.

The question, however, of whether place is one of the four kinds of cause, and in particular whether it is a final cause, is somewhat more difficult. For even though a bodily surface would hardly seem to have this status, the natural motion of the elements to their proper places and their natural rest once they arrive there might suggest to the contrary that at least being in such a place, if not strictly speaking the place itself, is that for the sake of which this natural motion exists.[86] And we recall that one of the requirements by which a definition of place was to be judged was its consistency with the assumption that there are natural motions of light and heavy bodies toward their proper places.[87] Accordingly, Aristotle continues his discussion with an attempt to show that it is reasonable that each thing should move toward and remain in its proper place. He bases his argument on the kinship between each element and the one that surrounds it, and he even likens the tendency of a lower element to remain in its place beneath the one above it to the tendency of a portion of one element to remain within the whole. In support of this striking suggestion, he makes the even more striking suggestion that the (element) in its place is a kind of part in relation to a whole. For since water, he continues, is potentially air (i.e., by evaporation), it is related to the air that surrounds it as material is to its being at work. And on the assumption, then, that the same water is in a sense air, as well as water, it would be related to air as a kind of part to a whole. This is the reason, he adds, that there is contact between an element and the one above it, whereas there is natural fusion when they both become one in the full sense (ἐνεργείᾳ ἕν), for example, when water is transformed into air.[88] Now it is not worth dwelling at length here on the details of this argument, since Aristotle acknowledges that he has only touched upon it unclearly, and he says that he will have occasion to develop it more clearly elsewhere.[89] But several points are worth noting. First, this argument is not based on Aristotle's explicit definition of place as the extremity of a surrounding body, but it corresponds instead to his suggestion that place is often better understood as a bodily being or bodily region as a whole. For that in which the element water tends to remain, by this account, is not the lower extremity of the surrounding air, but rather the entire region made up of the bulk of both of these elements. We should not, however, be surprised, at this stage of our interpretation, that no single definition of place can meet all the requirements of his original attempt to explain it in the "most beautiful" way. A second feature of this account that is worth noting here is its suggestion that the natural tendency of a body is not simply to move toward and to remain in its proper place, but also to fulfill its remaining potency by being transformed into the higher element. And on this basis, we can better understand why Aristotle did not explicitly claim to have resolved the perplexity that had denied that place is a (final) cause. For he surely knew that it is hardly a reasonable assertion to say that the natural motion of water exists in order for it to come to a place where it can become air—as if it would be better for water, or for the world as a whole, that this lower element cease to exist.[90] And more generally, he understood that to the extent that there are final causes of the motions of the elements, they are not evident unless one focuses on the contributions these motions make to the welfare of living beings, and especially man.[91]

Now that he has offered his definition of place, responded to most of the perplexities about it, and shown, to the extent possible, how the attributes it is believed to possess are indeed attributes of what he has defined, Aristotle does not continue to try to fulfill the remaining goal of his program of inquiry, which was to make manifest the cause of our discomfort and of the perplexities about place.[92] But he has already touched upon this question with his account of our illusions about matter and of our corresponding belief that place is "something great and hard to grasp."[93] And he has made it clear, I think, that the chief cause of all these illusions and of the ensuing difficulties is our wish to believe in place as something more than it is. Aristotle does not, however, address this point explicitly, since to do so in an adequate way he would have to acknowledge the extent to which his own definition of place, and especially the version that insisted on its being immoveable, has made concessions to this very wish. Instead, therefore, he simply concludes his whole discussion with the statement that he has said both that place exists and what it is.[94] But even this conclusion is less complete than we would have expected. For if we compare it to the beginning of this discussion, we see that Aristotle omits any reference to the third of his initial questions, namely, the question of the manner in which place exists (or "how it exists," πῶς ἔστι, 213a12–14). And he calls attention to this omission in the immediate sequel, where he begins his inquiry regarding void by asking not only whether it exists and what it is, but also about the manner of its existence.[95] It is true that later in this discussion of void Aristotle does say that he has already said "both how place exists and how it does not exist."⁹⁶ But this later statement only serves to confirm that its absence from the conclusion to the discussion of place was not a mere oversight. By stating belatedly that he has answered the question of how place exists (and of how it does not), but by not making this statement in the appropriate context, Aristotle is suggesting, I think, that he has both answered it and not answered it, or that he has answered it between the lines. For what he has shown is not merely what he says openly—which is that place depends on body and that it is not an independently existing interval—but also that it is something far less than we had expected it to be. He has shown, if we have followed his argument, that there is nothing that corresponds to all of our assumptions about place, let alone to all of our wishes, and hence nothing that could be explained in that "most beautiful" way in which he had encouraged us to try to explain it. But even this negative conclusion is an important step in our education about nature.

NOTES

1. *Physics* 209a29–30.
2. *Physics* 210b32–34.
3. *Physics* 211a6–11.
4. *Physics* 208a27–29.
5. *Physics* 208b1–209a2; cf. 205b31–34.
6. Cf. *Metaphysics* 1014a26–35; *On the Heaven* 306a9–11.
7. *Physics* 209a2–30; cf. 216b2–16.
8. One indication that Aristotle does indeed regard this as the common view of place is that he accepts it himself as the basis for his own treatment of the subject in the *Categories* (5a8–14).
9. Cf. Simplicius, *In Aristotelis Physicorum Libros Quattuor Priores Commentaria*, in *Commentaria in Aristotelem Graeca*, vol. 9, ed. H. Diels (Berlin, 1882), 537.32–538.14. The section on place from Simplicius's commentary on the *Physics* has been translated by J. O. Urmson in *Simplicius: On Aristotle's* Physics *4.1–5, 10–14* (Ithaca: Cornell University Press, 1992).
10. *Physics* 209a31–b17; cf. 207a21–26, 209b33–210a2; Plato, *Timaeus* 51a7–b1, 52a8ff.
11. See, however, p. 105 and chapter 5, pp. 121–24.
12. *Physics* 209b27–30.

13. *Physics* 214a14–15; *On Coming into Being and Perishing* 320b16–17; cf. Alexander of Aphrodisias, as quoted in Simplicius, *In Libros Quattuor Priores Commentaria*, 544.20–545.2.

14. *Physics* 191a8–12, 217a21–b11; *On Coming into Being and Perishing* 329a24–35. One might try to escape the difficulty referred to in the text by claiming that when Aristotle speaks of matter as inseparable from the thing, he means only that a thing's matter is distinguished from its place by being inseparable from it as long as it remains in existence (cf. Simplicius, *In Libros Quattuor Priores Commentaria*, 543.4–8). Yet though this interpretation would simplify the argument, it depends on the assumption that immediately after telling us that the question of matter requires "the keenest scrutiny," Aristotle would speak about it without saying quite what he meant. Also, there are a number of details in his ensuing discussion that are best explained by taking his remark here at its face value (cf. *Physics* 210a9–11 and pp. 83–86, 92–95).

15. *Physics* 209b30–210a9.

16. *Physics* 212a1–2.

17. *Physics* 210b31; cf. 210a11–13.

18. Compare *Physics* 210b27–31 with 209b28–30.

19. Cf. *On the Soul* 412a17–21, b6–9.

20. *Physics* 210a25–b4.

21. Compare *Physics* 193a9–12 and *Metaphysics* 1014b26–32 with *Metaphysics* 1017a5–6.

22. *Physics* 210a17–21.

23. Cf. *On the Soul* 412a19–b22 and see above, pp. 20–21.

24. *On the Heaven* 312a17–33 and, in particular, lines 30–31. It is true that Aristotle also says here that the elements have a single common matter (though its "to be" is different for each element). But this is largely a rhetorical concession, I think, to those who wish to believe in an unchanging substrate. There is also a second (rhetorical) reason for his not simply denying there that the elements have a single matter, a reason which I will discuss in the next chapter (see pp. 130–31 and n. 41). As for the difficulty of distinguishing form and matter in the four elements, see *Meteorologica* 389b28–390b2.

25. The case that Aristotle did not believe in any such "prime matter" of the elements has been well made by W. Charlton, in the appendix to his translation of *Aristotle's* Physics, *Books I and II* (Oxford: Oxford University Press, Clarendon Press, 1970), 129–45 (see also chapter 1, n. 21). I am here disregarding the sense of the term *matter* according to which water, for instance, since it can be transformed into air, is spoken of as being matter for it. Cf. *Physics* 213a2–4.

26. *Physics* 210b34–211a6; cf. 211a28, a33.

27. *Physics* 211a12–23.
28. *Physics* 211a23–b5.
29. *Physics* 212a2–6; cf. 211b6–9.
30. An additional wrinkle to this argument is that Aristotle presupposes the truth of his own proposed definition of place in order to reject the alternative that place is form. And yet he does, nevertheless, go on to rest the case for his proposal on the elimination of the three other alternatives. This inconsistency in his procedure serves to highlight the question of what, if anything, is the true basis for his proposed definition. Compare *Aristotle's* Physics, *Books III and IV*, trans. Edward Hussey (Oxford: Clarendon Press, 1993), 115.
31. Cf. Simplicius, *In Libros Quattuor Priores Commentaria*, 569.35–570.15; Simplicius *In Aristotelis Physicorum Libros Quattuor Posteriores Commentaria*, in *Commentaria in Aristotelem Graeca*, vol. 10, ed. H. Diels (Berlin, 1895), 868.25–871.15. See also Hussey, *Aristotle's* Physics, 114.
32. Cf. Simplicius, *In Libros Quattuor Priores Commentaria*, 604.33–605.5; Philoponus, *In Aristotelis Physicorum Libros Quinque Posteriores Commentaria*, in *Commentaria in Aristotelem Graeca*, vol. 17, ed. H. Vitelli (Berlin, 1888), 564.3–14; H. R. King, "Aristotle's Theory of ΤΟΠΟΣ," *Classical Quarterly* 44 (1950): 87–88.
33. Cf. *Physics* 212a18–19.
34. *Physics* 211b7–8, b14–20, 212a3–5.
35. *Physics* 211b5–14; cf. n. 30.
36. *Physics* 211b14–17.
37. *Physics* 211b18–23.
38. Cf. *Physics* 206a9–207b15. As for the difficulties that Aristotle was aware of even with regard to his own doctrine of the potentially infinite, see chapter 3.
39. Cf. Thomas Aquinas, *In Octo Libros Physicorum Aristotelis Commentaria*, ed. P. M. Maggiolo, (Rome: Marietti, 1965), bk. 4, lecture 6, para. 461. Aquinas's commentary has been translated by R. Blackwell, R. Spath, and W. E. Thirlkel as *Commentary on Aristotle's* Physics (New Haven: Yale University Press, 1963). See also "Aristotle and Other Pre-modern Thinkers on the Existence of Vacua," a doctoral dissertation by R. Glen Coughlin (Université Laval, 1986), 238–46.
40. Cf. Simplicius, *In Libros Quattuor Priores Commentaria*, 578.2–13, 621.20–30; Philoponus, *In Libros Quinque Posteriores Commentaria*, 562.1–563.2.
41. This position would also, of course, compel those who believe in the heliocentric hypothesis to deny that anything on earth can remain in the same place.

42. *Physics* 211b25–29.

43. Cf. *Physics* 210b22–27, 212b27–29.

44. Compare the notions of "absolute space" and "relative space" in Newtonian physics. Newton, *Principia*, vol. 1, *The Motion of Bodies*, Motte's translation revised by Florian Cajori (Berkeley and Los Angeles: University of California Press, 1934), 6.

45. *Physics* 211b31–212a2; cf. Joe Sachs, *Aristotle's* Physics: *A Guided Study* (New Brunswick: Rutgers University Press, 1995), 105.

46. Aristotle suggests, at all events, that there is *some* link between the illusion of an unchanging substrate and the common view of place as an independently existing interval. He points to the similarity between these two illusions by discussing the suggestion that place is matter only after the one that it is an independent interval, whereas his preliminary list of alternatives had mentioned these two in the reverse order (*Physics* 211b6–212a2).

47. Cf. *Physics* 209b17–21.

48. *Physics* 212a10–11.

49. *Physics* 209b28–30, 210a24, 211a34–b5; cf. 210b27–30, 211b25–29, 212a13–14.

50. *Physics* 212a20–21.

51. Cf. *Aristotle's Physics: A Revised Text with Introduction and Commentary*, ed. W. D. Ross (Oxford: Clarendon Press, 1936), 575–76; Richard Sorabji, *Matter, Space, and Motion: Theories in Antiquity and Their Sequel* (Ithaca: Cornell University Press, 1988), 188ff.; Hussey, *Aristotle's* Physics, 117–18. Sorabji gives adequate reasons, I think, for rejecting Ross's view that this definition abandons the requirement of contiguity between the place and the thing in place, or that it treats place as "the nearest unmoved boundary of a container, the first you would come to in moving outwards from the thing." Cf. *Physics* 212a29–30.

52. Cf. *Politics* 1276a34–b1.

53. Aquinas's suggestion that the extremities of surrounding bodies are immoveable places, despite the motions of those bodies themselves, by virtue of preserving a fixed "order or position" [*ordinem vel situm*] in relation to the immoveable world as a whole, fails to take account of this difficulty. On the other hand, Aquinas is right, I think, to suggest that the sameness—such as it is—of a given place is like the sameness of form in a being whose matter is constantly changing (Aquinas, *In Octo Libros Commentaria*, bk. 4, lecture 6, paras. 468–69; cf. G. Coughlin, "The Immobility of Place in Aristotle," *Philosophia Perennis* 2 [1994]: 13–17). This kinship between the (relative) permanence of place and the (relative) permanence of form in an enmattered being is a further reason

for Aristotle's emphasis in this discussion on the views of place as form or as matter.

54. Consider *Physics* 216a26–33 and context.

55. See, again, *Physics* 209a33–b1, which includes the only example in the *Physics* of the second person singular pronoun, i.e., Aristotle's only direct address to each of his readers.

56. *Physics* 209a31–b2; cf. p. 81.

57. Aquinas's claim that the whole river is only the common place of the moving boat, and that its proper place (at each instant) is the "order or position" of the flowing water in relation to the whole river, has no basis in Aristotle's text. Aquinas, *In Octo Libros Commentaria*, bk. 4, lecture 6, para. 468.

58. Cf. *Physics* 263a23–b9, and see also chapter 3, pp. 53–54.

59. *Physics* 211b25–28; cf. p. 92.

60. By "the vessel as a whole" I mean the containing body together with all its contents. Just as a river remains the same river despite the replacement of all the water in it, so a vessel, in this sense, remains the same vessel though all the fluid in it may change. Aristotle points to this view of the matter by referring, in his discussion of this example, to the vessel's contents as being parts of the whole vessel (*Physics* 211b25–26; contrast 211a29–b5).

61. From here we can perhaps understand Aristotle's odd expression that the fluids moving in the moving vessel do so in the place in which they are, "and not in the place in which they come to be, which is a part of the place that is the place of the whole world" (*Physics* 211b28–29). For the contents of the moving vessel do (by concomitance) come to be in a new place, or in a new region of the surrounding air, which place is a part of the larger place that is the air as a whole, and ultimately of the largest place that is the world as a whole. The claim, in other words, that a place is also a part of a larger place does not merely express the view of those who think of place as an independent interval. Their mistake is to imagine that place in the largest sense is something that contains the whole world, whereas in fact it consists of the whole world, and is the common place of every body (cf. *Physics* 209a32, and see also 210a17).

62. Cf. Simplicius, *In Libros Quattuor Priores Commentaria*, 585.34–587.16, especially 586.26–30.

63. *Physics* 212a21–30; consider, however, the reading of GIJ¹PST at 212a30.

64. Cf. Themistius, *In Aristotelis Physica Paraphrasis*, in *Commentaria in Aristotelem Graeca*, vol. 5, pt. 2, ed. H. Schenkl (Berlin, 1900), 119.11–12; Philoponus, *In Libros Quinque Posteriores Commentaria*, 587.16–21, 591.14–25; Hussey, *Aristotle's* Physics, 118.

65. Cf. *On the Heaven* 278b11–15.
66. *Physics* 212a35–b3.
67. Cf. *Physics* 221b12–14.
68. Simplicius, *In Libros Quattuor Priores Commentaria*, 589.31–590.22.
69. Cf. *Physics* 210a26–30, 224a23–26.
70. Cf. *Physics* 211a12–14.
71. That Aristotle regards circular motion as a kind of locomotion is clear from *Physics* 261b27–29 (cf. 261a27–28) and *On the Heaven* 268b17–18, among other passages. Cf. Simplicius, *In Libros Quattuor Priores Commentaria*, 602.8–603.22.
72. Cf. *Physics* 254b8–10; contrast 224a21–28.
73. Cf. Themistius, *Paraphrasis*, 121.1–9. An interpretation suggested by some commentators is that all the parts of the outermost sphere itself are "in a sense," i.e., potentially, in place, since they all surround one another, i.e., succeed one another in the circular direction, though they are not actually divided from one another (cf. Alexander of Aphrodisias as reported by Simplicius, *In Libros Quattuor Priores Commentaria*, 593.7–23; Aquinas, *In Octo Libros Commentaria*, bk. 4, lecture 7, para. 484; and see *Physics* 212b3–4; 211a17–21, a29–31). This interpretation does give some significance to Aristotle's distinction here between being potentially and being actually in place, and it is also consistent with the language of the text (ἐπὶ τῷ κύκλῳ γὰρ περιέχει ἄλλο ἄλλο, 212b13, emphasis mine). Yet this view of the place of the parts of the outermost sphere fails to make any sense of their motion. For these succeeding parts maintain the same relations to one another—and thus remain in the same "places," by this account—whether the sphere is in motion or at rest.
74. Cf. *On the Heaven* 310b7–14; *Physics* 209a33–b1.
75. This suggestion about the surface of the earth provides some limited justification, I think, for Ross and those other scholars who interpret the "first" immoveable limit, as it is referred to in Aristotle's final definition of place, as the *nearest* unmoved boundary of a surrounding body, even though it may not be contiguous with the one in place. See Ross, *Aristotle's* Physics, 575 and n. 16; cf. F. M. Cornford, in *Physics, Books I–IV*, ed. and trans. P. H. Wicksteed and F. M. Cornford, Loeb Classical Library (1929), 314.
76. Cf. *Posterior Analytics* 89b29–31.
77. Aristotle's arguments in *On the Heaven* (296a24–b26) that the earth remains unmoved at the center of the world—arguments that presuppose the interpretation that I am here challenging of what it means for it to remain unmoved—can best be explained, I think, as further con-

cessions to the popular understanding. (Consider the critique of these arguments in Galileo, *Dialogue Concerning the Two Chief World Systems*, 2nd ed., trans. Stillman Drake [Berkeley and Los Angeles: University of California Press, 1967], 32–36, 124–41.)

78. *Physics* 212b13–22.

79. This suggestion is supported by the fact that most of the manuscripts and one of the ancient commentators add to the phrase that refers to this extremity two additional words meaning "a limit that is at rest" (cf. *Physics* 212b19–20).

80. Cf. *On the Heaven* 270b16–24.

81. E.g., Themistius, *Paraphrasis*, 121.15–20; Simplicius, *In Libros Quattuor Priores Commentaria*, 594.7–27; Philoponus, *In Libros Quinque Posteriores Commentaria*, 604.9–16; Aquinas, *In Octo Libros Commentaria*, bk. 4, lecture 7, para. 485; Ross, *Aristotle's* Physics, 578.

82. Cf. *On the Heaven* 270b24–25, 302a28–b5.

83. The suggestion that each place of the growing body exists only instantaneously, and is thus incapable of either motion or rest, is hardly a convincing alternative. See Sorabji, *Matter, Space, and Motion*, 188–89.

84. Cf. *Physics* 210b22–27.

85. *Physics* 212b22–29.

86. Cf. *Physics* 208b8–22; Simplicius, *In Libros Quattuor Priores Commentaria*, 533.19–25.

87. *Physics* 211a3–6; cf. p. 87.

88. *Physics* 212b29–213a10.

89. Aristotle indicates that he will have to explain, for instance, why water is the material for air in a different, and presumably truer, sense than air is the material for water. Cf. *On the Heaven* 310a31–b15ff., 312a12–21; *On Coming into Being and Perishing* 318b1–33, 335a14–21.

90. Cf. *Physics* 198b8–9; and see Simplicius, *In Libros Quattuor Priores Commentaria*, 600.4–6ff., 606.16–20.

91. Cf. *Physics* 194a34–36, 198b17–199a8; and see above, p. 47 and n. 37.

92. Cf. *Physics* 211a9–11.

93. *Physics* 212a7–8; cf. 211b31–36; and see pp. 92–95.

94. *Physics* 213a10–11.

95. *Physics* 213a12–14.

96. *Physics* 214a17–18.

CHAPTER 5

The Doctrine of Weight and Lightness

The discussion of weight and lightness with which Aristotle concludes his *On the Heaven* (Περὶ Οὐρανοῦ) is a most appropriate ending to the work. For even if these attributes, and the natural motions to which they give rise, are not simply the primary characteristics of the sublunar elements, they are the ones responsible for their separation into distinct places, and thus they are largely responsible for there being a world (οὐρανός), or an ordered whole, at all.[1] Moreover, Aristotle has relied on the distinction between heavy bodies, which move downward by nature, and light ones, which move naturally upward, as an important factor in his earlier discussions. He appealed to it, with a promise of a more precise treatment later, as a key premise in his argument for the existence of a distinct kind of heavenly body that moves naturally in a circle.[2] And as for the sublunar region, a major part of his evidence that there are such things as simple or elemental bodies, and that they are neither infinitely many nor only one in kind, is that there are simple motions characteristic of this region and that these motions, the motions upward and downward, are themselves neither infinitely many nor only one.[3] And yet Aristotle acknowledges that the hypothesis of these two simple motions is a controversial one. For he notes that most, if not all, of his predecessors in natural philosophy regarded the bodies that he calls naturally light as being merely less heavy than the others, and he refers to an account that attributes their upward motion to their being forced or squeezed upward by the greater weight of the surrounding body.[4] Thus, it is incumbent on him, for the sake of his whole teaching in *On the Heaven*, to defend his claim that upward and downward motions are equally fundamental; and he does this by concluding the work with the more precise treatment of the light and the heavy that he had promised.[5]

Aristotle begins this new discussion with the claim that the heavy and the light are spoken of both simply, or absolutely, and in relation to another, or relatively. And after asserting that his predecessors have spoken only of the relative weight and lightness of bodies that have weight, he turns to his own definition of what the heavy and the light are in themselves. To prepare this definition, he first explains what he means by up and down, or above and below. Unlike the many, who falsely believe that the hemisphere above their heads is the only heaven, and hence that up and down each refer to a single definite direction, Aristotle is aware that the same heaven surrounds both hemispheres of the earth. And yet he supports the many, against the claims of their more sophisticated critics, in their belief that the up and the down are definite parts of the world, and even in their belief that the up is "higher" than the down, in dignity as well as in place. He does this by redefining the up as the entire region away from the center, or at the extremity of the world—which region he speaks of as being both above in position and first by nature—and by a corresponding redefinition of the down as the region at the center. Having thus clarified what he means by up and down, he proceeds to define the simply light as that whose (natural) motion is (always) upward, or toward the extremity of the world, and the simply heavy as that whose (natural) motion is (always) downward, or toward the center. As for the relatively light, or the lighter, he defines it as that one of two bodies, both of which have weight and which are equal in volume, whose natural motion downward is exceeded in speed by that of the other.[6]

These definitions of the simply heavy and the simply light are straightforward enough, though one may of course doubt whether there are such bodies, and even whether the world has a center and an extremity, as Aristotle claims. But the definition of relative lightness, which speaks only about the relationship between bodies of equal volume, is not immediately clear. Surely, it cannot be a complete definition. Aristotle himself will say a bit later that a larger body that has weight moves downward more quickly than a smaller one of the same kind, thus implying that this smaller body is lighter than the larger one, as of course it is.[7] What he must mean, then, in defining relative lightness in the way he does is that one *kind* of body is lighter than another kind if its natural downward motion is slower than that of an equal volume of the other.[8] But even on this understanding, the specification of equal volumes is somewhat puzzling, since Aristotle will later claim that when one element is lighter than another, even if its lightness is only relative, no amount of it, however great, is heavier than any amount of the other, however small. As evidence for this claim, he appeals to the fact that any amount of the lighter element rises to the surface of the heavier one, just as any amount of the heavier one sinks to the bottom of the lighter one. And this view of the elements is so central to his overall argument that one of his chief objections to the notion that all bodies are composed of a common, heavy material is that it would imply that a sufficiently large amount of a lighter element is heavier than a sufficiently small amount of a heavier one.[9] But if Aristotle regards this consequence as being untenable, why, to repeat, does he first define the relatively light in comparison only to an equal volume of a heavier body?

This question about the definition of relative lightness will turn out to be an important one; but rather than try to answer it at once, let us keep it in mind as we follow the course of Aristotle's own argument, beginning with his examination of the views of his predecessors. He introduces this examination with a restatement of his general criticism that others have spoken only of the relatively light and heavy, while failing to define the simply light and the simply heavy. To illustrate what he means by this criticism, he refers to the doctrine that all bodies are composed of the same kind of parts, such as the elemental triangles of Plato's *Timaeus*, and that those which have more of these parts are heavier and that those which have fewer are lighter.[10] His contention is that this doctrine cannot explain the motion of what is simply light, i.e., fire, which he says is always light and always moves upward. (One should note that according to Aristotle the element "fire" in its purity is not the fire that we see burning around us, but rather a distinct substance—like a kind of hot and dry smoke—whose proper place is above the air and below the innermost sphere of the heaven.[11]) If this doctrine were true, he says, a greater quantity of fire would be heavier than a smaller one, and would move upward less readily, since it would possess more of the same parts; but he objects that in fact, the greater the quantity of fire, the lighter it is and the more quickly it rises. Besides, he adds, since (according to the *Timaeus*) air, water, and fire are made up of the same triangles, there would have to be some amount of air that is heavier than (some amount of) water. But he claims that what actually happens is entirely contrary to this view; for any amount of air rises upward from water, and the more there is, the more it rises.

Having stated his objections to this first doctrine, Aristotle turns to another one, which he regards as being superior to it, at least in some respects. For the view that bodies are heavier than others by being made up of more of the same parts fails to account for the fact that a heavier body is not necessarily greater in bulk than a lighter one. And to meet this difficulty, some earlier thinkers had thought of another cause and said (οἴονταί τε καὶ λέγουσιν, 309a5–6) that there is void within bodies, which makes them lighter. Accordingly, they said, larger bodies are sometimes lighter than smaller ones, since they contain more void, and therefore often only an equal number, or even fewer, of (the primary) solids.[12]

Aristotle acknowledges that this hypothesis of a void within bodies allows for an explanation of the fact that larger bodies are sometimes lighter than smaller ones. But he objects that it too can not explain why some bodies should be simply light or simply heavy, and thus always move upward or downward. For if, he says, the lightness of fire is accounted for by the larger amount of void, and (hence) the smaller amount of solid, that it contains, there will be some quantity of fire in which the amount of solid surpasses that contained in a small quantity of earth. And if the proponents of this doctrine should reply that such a large quantity of fire would also contain more void—with the implication that this void is in itself the cause of lightness—Aristotle asks in return how they are going to define the simply heavy. It must, he says, be either by its having more solid or else less void. But if they say it is that which contains more solid, Aristotle's response is that there would have to be some quantity of the simply heavy element, earth, so small that it contains less solid than a large quantity of fire. And if they say it is that which contains less void, then it follows that (a large quantity of) something that always moves downward can be lighter than (a small quantity of) that which is simply light and always moves upward. But this, Aristotle argues, is impossible, on the grounds that no matter what the quantities, the simply light is always lighter than the bodies that have weight and move downward. Nor, he continues, does it resolve this difficulty to suggest that the ratio of the void to the solid within a body is the cause of its lightness or its weight. For although this hypothesis does indeed offer an account of why any amount of fire is lighter than any amount of earth, it contradicts the obvious fact that different amounts of a given kind of body do not all have the same weight or lightness. A large amount of fire, for instance, would have the same ratio of void to solid as a smaller one, and thus the same degree of lightness, according to this view; yet the fact is that it moves upward more quickly—a sign, as Aristotle presents it, of its being lighter—just as a larger amount of gold or lead moves downward more quickly than a smaller one.[13] Thus, it seems, the notion of a void within bodies, no matter how it might be elaborated, is inconsistent with the appearances of motion.

Given the fact, however, that Aristotle made a suggestion of his own to help save the doctrine of an interstitial void, namely, the suggestion that the ratio of void to solid within a body might be the cause of its lightness or its weight, the question arises of why he did not also suggest the more plausible view that this cause might instead be the *difference* between the body's void and its solid.[14] An account of lightness and weight in terms of this difference, or in other words, in terms of the excess of the amount of void in the body over the amount of solid, or vice versa, could have avoided the objections Aristotle made against the alternatives that he did mention. For it leads to the consequence that any amount of fire, for instance, if fire contains more void than solid, is lighter than any amount of earth, assuming that earth contains more solid than void. And at the same time, it leads to the consequence that a larger amount of a light body is lighter—and of a heavy body, heavier—than a smaller one. To be sure, there may be other valid objections to this suggestion, and in fact there are, as we shall see. Yet its adequacy for defending the doctrine of void against the objections that Aristotle himself has brought forward in this discussion does lead one to wonder why he did not even mention it.

Now this question, like the earlier one about Aristotle's definition of the relatively light, is one that it is better not to try to answer at once. But it is worth noting that in the sequel he outlines another objection to the doctrine of interstitial void and that this objection, which seems otherwise superfluous, is precisely suited to refute the attempt to explain lightness and weight in terms of the difference between the void and the solid within a body. It thus adds credence to the possibility that Aristotle is thinking of this alternative and that our question about his failure to mention it is one that he wishes for us to ask (cf. n. 14). The new objection is that the doctrine of void and solid cannot explain why the intermediates between the simply heavy and the simply light are heavier and lighter than each other as well as the (two) simple (elements). What this compressed expression seems to mean is that the doctrine cannot explain why air, the lighter of the two intermediates, is always lighter than the other one, water, no matter what the amounts of each, while at the same time always being heavier than the simply light element, fire, and that it likewise cannot explain the corresponding facts about water in its relation to air and to earth.[15] Now this argument, to repeat, would seem to be unnecessary against the already rejected alternatives that only the solid or else only the void within a body is directly responsible for its lightness or its weight, and it is not even valid against the view that these are caused by the ratio between void and solid. But it does make sense as an objection against the attempt to explain these qualities in terms of the difference, rather than the ratio, between void and solid. For although such an account can perhaps explain why air is always lighter than water—if air contains an excess of void, and water an excess of solid—it does so at the cost of allowing that some large amount of air, with a correspondingly large excess of void, will be lighter than a smaller amount of fire and, likewise, that some large amount of water will be heavier than some small amount of earth. Still, despite the fact that this objection is so appropriate against the attempt to explain lightness or weight in terms of the difference between the void and the solid within a body, Aristotle presents it as an objection to the doctrine of void and solid in general, and he makes no explicit reference, either here or elsewhere, to this particular interpretation of it.

In this discussion of the views of his predecessors, Aristotle makes some additional arguments against the doctrine of void and solid. He asks, for instance, how void can be the cause of upward motion in bodies without moving upward itself; yet he also questions whether the notion of a void that does move upward, and that nevertheless remains together with the downward moving solid component in bodies, is even intelligible. And he goes on to suggest, if only implicitly, that those proponents of an interstitial void who interpret it as being the mere absence of solid are offering a version of the previously rejected view that all bodies are made up of a single, heavy material.[16] But what remains most important throughout this whole discussion is that Aristotle presents the various doctrines as sharing a common weakness. For whether they posit a single material or a pair of contraries, they all imply that some amount of at least one element that is lighter than another will be heavier than some other amount of the heavier one. And as we have seen, Aristotle asserts that this is impossible.

Having prepared the way with this critique of the views of his predecessors, Aristotle turns to his own account of lightness and weight, which begins from the general question of why each kind of body moves to its own place. His response to this question is based on the premise that a body's motion to its own place is the natural completion of its coming into being, or of its change into its own form as something heavy or light. What he supposes, in other words, is that just as a child that has not reached maturity is not yet fully itself as a human being, so also a newly generated body that has not reached its own place is not yet fully itself.[17] And in accordance with this premise, he suggests that the cause of the body's motion toward this place of its own is an inner tendency toward the attainment of form. Thus, for instance, when water evaporates into air, the newly formed air tends upward by its own nature to its own place, or to the actuality of its form as something light.[18]

This simple answer, however, to the question of why bodies move to their own places is not the whole story. For rather than appealing directly to a natural tendency toward the attainment of form, Aristotle claims instead that a body's motion to its own place—or its own form, as he has just supposed—is no more problematic than are healing and growth in the case of beings with the potency for these motions. And yet he goes on to observe that these latter motions are thought to depend on external causes, whereas the natural locomotion of light or heavy bodies is thought to originate from within. Thus, the attempt to account for this motion on the analogy with these others seems inadequate. But Aristotle defends his analogy by replying that growth and healing sometimes also proceed on their own after only a small external movement.[19] He still acknowledges that the heavy and the light appear to contain their own principle of motion *to a greater extent* than does that which can be healed or that which can grow. But later in his discussion, he minimizes even this apparent difference. For on the one hand, he says that the motion of the elements depends itself on the prior action of whatever it was that first brought them into being; and on the other hand, he suggests that the external movements necessary for growth or for spontaneous healing are merely the removal of hindrances, hindrances analogous to those that can prevent an already formed element from moving to its own place. Thus, not only is there an external moving cause in all three kinds of motion under discussion, but in all these cases, this external cause seems to be less immediately relevant than the internal one. However, the manner in which Aristotle presents this latter suggestion raises doubts about his treatment of external causes. For rather than saying, as we would have expected, that it is the being with the capacity for growth that needs only the removal of hindrances in order to begin its motion, he makes this claim about the nourishment from which it grows.[20] Yet if the transformation of nourishment into being a part of what feeds on it can be interpreted as its own motion, the distinction between motion from an internal principle and that whose primary cause is external seems tenuous indeed. And the question thus arises of whether the blurring of this distinction, in the case of nourishment and its transformation, is meant to suggest any parallel in the case of the motion of the elements to their own places.

A further difficulty with Aristotle's general account of why things move to their own places is the dubiousness of its premise that this motion is a motion toward their own forms as heavy or light bodies. This premise has come, moreover, as a surprise, since

Aristotle had said earlier that things are called heavy or light by their *potency* for natural motion (τῷ δύνασθαι κινεῖσθαι φυσικῶς πως, 307b31–32) and had suggested that the fulfillment of this potency is their impetus (ῥοπή, cf. 307b33) upward or downward, rather than their having reached a certain place. He had even said, in a still earlier passage, that heavy bodies retain their weight or their propensity downward until they reach the center (of the world), with the clear implication that they would no longer be heavy, in the full sense, if they ever arrived at this place of their own.[21] But at all events, he does now offer the antithetical premise that it is only in their own places that heavy or light bodies are fully themselves, and to explain it he appeals to his definition of a body's place as the limit of that which surrounds it.[22] He goes on to claim that all things that move upward or downward are surrounded by the extremity (i.e., the inner surface of the heavenly sphere) and the center, and that these places are in a certain manner the form of what they surround. As for the intermediate bodies, water and air, which do not naturally reach the extremity or the center, he says that their places, or their forms, are the bodies consecutive to them in the cosmic order. He even interprets the old saying that "like moves to like" as meaning not that bodies move toward others of the same kind, but that they move toward (the place constituted by) these neighboring bodies, which they also resemble, at least in some respects.[23] Now in asserting that the surrounding bodies are a kind of form of the one they surround, Aristotle is implicitly treating this latter one as a kind of material for them both. In the sequel, however, he modifies somewhat this picture of the relationship of form and material among the elements. For he says that (only) the higher of the surrounding bodies is related to the one surrounded as form to material; and thus the whole sublunar sphere seems to belong to a single whole, with each body being the material for the one above it and, except in the case of the lowest body, the form for the one below it.[24] Yet despite the seeming coherence of this picture of the sublunar sphere, it remains unconvincing. For how can one body, or even its inner surface, be the form of another? Even if it were true that the inner surface of the higher body determined the contour of the one beneath it, and even if one could interpret the term *form* as a mere limiting surface, the surface of the containing body is not the same as that of the one contained.[25] Moreover, it is hard to think of a body's form in the sense of its contour as being the same as its form in the sense of its actual heaviness or lightness. And as we reflect on the difficulties in this elaboration of the premise that a body's motion to its own place is motion to its own form, we are prepared to notice that Aristotle has not even asserted that it is. For all he has said, in fact, is that *if* this premise is true, then a body's motion to its own place is analogous to the motion toward health or toward an increase in size.[26] And the dubiousness of this premise, taken together with the difficulty that has been suggested by these analogies, makes it clear that we have not yet seen an adequate account of why a body moves to its own place, or even of what it means for it to do so.

Aristotle's explicit claim, however, is that he has already answered these questions. Accordingly, he turns to a discussion of the differences among light and heavy bodies and tries to explain some of the various things that happen in their interactions with one another.[27] Yet as we shall see, these more particular discussions also offer important help with our general questions about a body's motion to its own place. Now to begin this new section of his account, Aristotle restates what he means by the terms *simply heavy* and *simply light*. The simply heavy, he now says, is that which sinks to the bottom of everything (else), and the simply light is that which rises to the surface of everything (else). Somewhat surprisingly, in the light of what he has reported earlier about his predecessors in natural philosophy, he claims that these definitions are in accord with what is apparent to everyone. But all he may mean by this claim is that everyone would agree that there is one kind of body that sinks below all the others and another kind that rises above them. And by calling these bodies simply heavy and simply light, and not merely heaviest and lightest, as he had done, for instance, at the beginning of *On the Heaven*, he does not yet raise any serious questions.[28] But what remains controversial is whether these heaviest bodies are simply heavy—and, even more so, whether these lightest ones are simply light—in the particular sense that he goes on to specify, namely, that they contain nothing of the opposite quality. Later, he will try to prove that this is so; but for now, he leaves it as a mere assertion. He also says that he is calling these bodies "simply" heavy and light with a view to their genus, in part, perhaps, in order to forestall the objection that a smaller amount of earth must be lighter, or a smaller amount of fire heavier, than a larger one. Aristotle turns next from the simply heavy and light to those bodies, such as water and air, that contain both weight and lightness, as is evidenced by their rising to the surface of some things and sinking to the bottom of others. These he calls heavy and light in a different sense. But although these intermediate bodies are not simply heavy or light, he claims that they are, respectively, simply heavy and simply light in relation to one another. For as he reminds us, any amount of air, however small, rises to the surface of water, and any amount of water sinks to the bottom of air.

Having prepared the way with this statement about the four elements, Aristotle goes on to discuss composite bodies, and he says that their differences in weight and lightness must be traced to the differing amounts of each of the elements that they contain. In particular, he says, this is the reason that the same bodies are not thought to be heavy and light in all places. As an example of what he means, he refers to the fact that a talent of wood is heavier in air than a mina (a sixtieth of a talent) of lead, whereas it is lighter in water. And his explanation of this anomaly is that wood contains more air than it does earth and water, and that this air, which has no tendency to rise when the wood is surrounded by air itself, does tend to rise, thus bringing the wood to the surface, in water.[29]

Despite the plausibility, however, of this explanation of the different attributes of wood in air and in water, the manner in which Aristotle presents it raises questions. For rather than presenting it immediately, as I have just done, he begins with the more general statement that all (the elements), except fire, have weight and that they all, except earth, have lightness. Thus, he continues, earth and all bodies in which earth predominates have weight everywhere, water has it everywhere except in earth, and air has it everywhere except in water and in earth. For even in their own places, he says, all the elements except fire have weight. And as a sign that this last claim is true even in the case of air, he says that an inflated bladder draws (a scale) downward more than an empty one does. Now this argument for the weight of air in its own place is the immediate preliminary from which Aristotle concludes that the air in a talent of wood is the cause of its being heavy in air, but light in water. Yet the weight of air in its own place hardly seems relevant toward explaining these facts. For the downward propensity of the air within wood is no greater than that of the surrounding air, which also has weight in its own place. And even if the air within the wood were somehow compressed, the amount of air in a talent of wood would not be a significant factor in its overall weight in air. For to take Aristotle's own example, an inflated bladder does not weigh significantly more in air than an "empty" one does, if indeed there is any noticeable difference at all.[30] Thus, the explanation for the anomaly that wood is heavy in air and yet light in water is not the weight of air in its own place. What is significant, rather, is its lightness in water. And indeed, Aristotle's final statement about the matter said only that bodies, like wood, that contain more air than they do earth and water will rise in water but not in air; and it did not even mention the fact that they also sink in air.

But why, then, does Aristotle give so much emphasis to the weight of air, and more generally the weight of all the elements except fire, in their own places? Let me suggest that he does so in order to prompt the reader to the following thought. Although the weight of air in its own place plays no significant role in explaining the weight of wood in air, the weight of *water* in its own place is the chief reason for the lightness of air, and therefore also of wood, in water. For the weight of the surrounding water, combined with the ease of its lateral movement, causes it to press in underneath any immersed body whose weight is less than that of water itself, thus creating an upward pressure that overpowers the downward propensity of the immersed body.[31]

Now this suggestion, although it is now acknowledged by all students of physics as being the true cause of buoyancy or lightness in fluids, is likely to seem incredible as an interpretation of the thought of Aristotle. For he has been emphatic in arguing that the upward motion of light bodies is as natural, and irreducible, as is the downward motion of heavy ones. But to this objection one could say at first that he may not have regarded the natural upward motions of the intermediate bodies, including air, as being so irreducible to motions in their surroundings as he did the upward motion of fire. And more importantly, his argument in the sequel, which claims to show that there is in fact a simply light body, contains obvious difficulties; and these difficulties, taken together with those that we have already discussed and also with others that we shall see later, point to the conclusion that he did not believe the whole of his own doctrine of the simply light. In order to confirm that this is so, let us continue with our analysis of the text.

In the following argument, which claims to show that there is a simply light body, as well as one that is simply heavy, Aristotle returns to his original definition of the simply light as that which naturally always moves upward, adding only the proviso that its motion not be hindered. Now for there to be a body that moves naturally upward, there must presumably be a place toward which it moves. Thus, Aristotle proceeds to argue that there is such a place, the extremity of the world, or at least the extremity of the region in which the four elements have their motion. This aspect of his argument is based on the premise that there is a definite center, to which, it seems, an extremity must be opposed. And on the analogy with earthy bodies, which in sinking to the bottom of all the others are agreed to be moving toward the center, he concludes that fire, which rises to the surface of all the others, must be moving toward the extremity.[32]

Aristotle's argument, however, for the existence of a simply light body does not end with this conclusion. For the argument as I have presented it so far does not even claim to exclude the possibility that fire might be forced upward toward the extremity, despite having a downward propensity of its own, as a result of the greater downward propensity of the surrounding medium. Accordingly, in order to respond to this denial of the irreducible character of its upward motion, he goes on to draw the further conclusion that fire has no weight at all. Having no weight, we recall, is an aspect of being simply light that Aristotle had specified in his recent discussion of the four elements (p. 124). And yet the basis from which he derives this further conclusion is clearly inadequate, as we shall see by looking at the argument in his own words. "If, then," he says, "there is something that rises to the surface of all things, as fire manifestly moves upward even in the air itself, although the air is [manifestly] at rest, it is clear that this [something] moves toward the extremity. So it is not possible for it to have any weight. For [then] it would sink to the bottom of something else. And if this [were the case], there would be something else that moves toward the extremity, which rises to the surface of all the things that move. But now there is nothing apparent. Therefore, fire has no weight . . ."[33] Now this conclusion that fire has no weight, as distinct from the prior one that it moves toward the extremity, explains Aristotle's introduction of the otherwise unnecessary premise that the air is at rest. For this is the premise that he needs in order legitimately to dismiss the claim that fire moves upward only because the downward propensity of the surrounding air is greater than its own. And yet in a context where he has just asserted that air has weight in its own place, it no longer makes sense that he should assume that this air, or rather the part of it that surrounds a lighter body, is at rest. Indeed, Aristotle knew, of course, that the air in immediate contact with an ascending body is never at rest, no matter what the cause of the body's upward motion.[34] Thus, his premise that the air is at rest must refer only to the air as a whole; and so he can hardly have imagined that he has excluded the possibility that fire is pushed upward, even though it has weight, by the air in immediate contact with it. And it should not come as a surprise, then, that he continues with his argument for the existence of the simply light and that, in his final statement of its conclusion, he repeats only the claim that there is something that moves toward the extremity, while saying nothing about its lacking all weight.[35]

A further sign that Aristotle did not think of even fire as lacking in weight occurs in a subsequent argument, despite the fact that he does repeat there his claim to the contrary.[36] The explicit aim of this new argument is to confirm his previous statement that the intermediate elements, as well as earth, have weight in their own places. As for its context, he has just asserted, in surprising agreement with those predecessors whom he had criticized for this very assertion, that the four elements all share a single, common material.[37] To be sure, he has insisted that this material has a different character, and even a different being (τὸ εἶναι, 312a20, 33), in each of the elements. But he does nevertheless assert that it is also one. And he does not spell out all his reasons for making this surprising claim. For though he mentions the fact that the four elements come into being from one another, he makes it clear (ἄλλως τε καὶ, 312a32) that this is not his initial evidence for their having a single material. On the other hand, in the course of his ensuing argument that the intermediate bodies have weight in their own places, he does, as we shall see, point indirectly to this missing evidence, at the same time as he hints at the truth that even fire has weight.

Aristotle's new evidence that the two intermediate elements have weight—and as he now adds, no lightness—in their own places is that they each move downward into the place of the body (naturally) beneath them whenever that body is removed. By contrast, he adds, air will not move upward to the place of fire, even if the fire above it is removed, except by violence, as also happens with water, which can be drawn upward (by violence) into a (heated) vessel. What he is referring to by this latter claim is the upward motion of water into a heated and inverted flask as the air within it cools. And he says that this water moves upward, not because it is light (which it is, as he has said, only in relation to earth), but rather because its surface becomes united with that of the air above it; the air then draws the water upward (σπάση τις ἄνω, 312b10) more quickly than it tends to move downward on its own. This account, however, of the upward motion of water in an inverted vessel does not seem to offer the expected analogy to Aristotle's suggestion that air could move upward by violence if the fire above it were removed. For what, in this latter case, corresponds to the gradually cooling air in the inverted vessel? Yet without such a body to unite its lower surface with the upper surface of the air, what would draw the air upward?[38] And by noticing this difficulty, we are better prepared to see another one, namely, that Aristotle's account of the water in the inverted vessel suffers from the same incoherence as to the agent of its motion. For if it is drawn upward by the enclosed air, then this air must presumably have an upward motion of its own (at least while it cools). But Aristotle has just told us that air has weight, and no lightness, in its own place, and that it has no upward motion of its own, except when it is rising within one of the heavier bodies. So why would the cooling air in the inverted vessel, even assuming that its surface becomes united with that of the water beneath it, draw the water upward? One is tempted, perhaps, to think that the air condenses as it cools, and that it rises upward, rather than downward, in order to avoid a vacuum within the vessel. But Aristotle never mentions this notorious doctrine of nature's so-called abhorrence of a vacuum in order to resolve this, or any other, question.[39] And yet in the light of what he does say, it seems incredible that this question could have escaped him. Mustn't he have known, then, that as the air in the vessel cools and thus loses some of its resistance to compression, it is *pushed* upward by the water beneath it, and that this water is in turn pushed upward as a result of the greater downward pressure of the surrounding air?[40] And mustn't at least part of his motive for referring to this phenomenon in the context that he does—namely, in an argument that air and water *both* have weight, but no lightness, in their own places—have been the wish to suggest this true explanation of it to his attentive readers? Moreover, if he knew, as I think he did, that the weight of air in its own place is thus responsible for the upward motion of water in an inverted vessel, mustn't he also have known that it is similarly responsible for the natural upward motion of fire, or of any body less heavy than the air itself? Mustn't he have known, in other words, that even the lightest of the sublunar elements also has weight?

Now it is true that Aristotle says here, as he has said earlier, that fire does not have weight, not even in its own place, and he gives this as the reason that it would not sink down into the place of air if the air were removed. But how can he claim to know that it would not do this? For while he could know from experience that air envelops the earth in the absence of water and that water (as well as air) sinks to the bottom of even the deepest pits, he could have had no experience of what happens at the upper surface of the air. Furthermore, his final explanation of why water and air sink below their own places when the elements beneath them are removed is the similarity of material among the three lower elements. And yet as we have noted, Aristotle has just claimed in this very context that all *four* of the (sublunar) elements have a common material. And he has pointedly failed to tell us the primary basis for his making this claim. Let me suggest then, in the light of my previous argument, that he in fact thought that even fire would sink downward to fill any places unoccupied by the lower elements. And let me also suggest that the primary basis for—and indeed the truest meaning of[41]—his claim that the four elements share a common material is that they have this common tendency, when unimpeded by a heavier body, to move downward toward the center of the earth.

There is, moreover, direct evidence from *On the Heaven* itself that Aristotle did think that even fire has weight. This evidence occurs in a much earlier argument, whose purpose is to show that our world is the only world.[42] Aristotle says there that there are three kinds of bodies, one that sinks toward the center, another that revolves in a circle, and an intermediate kind, whose place, he argues, is necessarily between the places of these others. For if it were not between them, he says, this intermediate body, which he now refers to as that which rises to the surface (of the body at the center), would have to be outside (of the places of these other bodies). In that case, of course, our world would not be the only world, since to be outside of the place of the body that revolves in a circle is to be outside of the heaven at the extremity of our world. But Aristotle goes on to argue that this situation is impossible, on the grounds that the place of a body with weight is lower than that of one without weight. Accordingly, he concludes that this third kind of body must be in the place between those of the other two.[43] The differences within this intermediate body itself, he adds, will be a matter for later discussion. Now the natural interpretation of this passage is that the intermediate kind of body, which rises to the surface of the one at the center and yet remains within our world, includes at least fire and air. And on this view, Aristotle is explaining that these intermediate bodies remain beneath the revolving heaven because they have weight, whereas it does not. In other words, the natural interpretation of this passage involves the claim that even fire has weight. Now since this claim contradicts what Aristotle explicitly says in so many other places, commentators have been at pains to find a different interpretation of this passage.[44] But I contend that its natural interpretation is the true one and that Aristotle is here admitting, almost in so many words, that fire, like the other three sublunar elements, remains as near as it does to the center of the earth because of its having weight.

To return, however, to the passage we have been discussing, Aristotle continues with his explicit denial that fire has weight. But a closer look at his argument shows again that he is only keeping up appearances, while offering still more indications of his true position. This next argument, which seems at first to be a mere restatement of his earlier critique of his predecessors, denies in general their assertion that all the elements have a single material, even though he had only recently agreed to this assertion himself.[45] His grounds for denying it are that if there were just one material, either a light one or a heavy one, then all things would move either upward or downward, and the other kind of motion would no longer exist. More precisely, as he continues, there would no longer be anything *simply* light if all things had (merely) a greater (or lesser) downward propensity by virtue of being made up of more (or less) of a single, heavy material. Yet he contends that we see this (i.e., the existence of the simply light), and he adds that it has been shown that (something) moves upward, as well as downward, everywhere and always. It should be clear by now, however, that Aristotle was aware that being simply light in *this* sense, namely, that of always rising to the surface of any other body (in the neighborhood of the earth), is not incompatible with having weight. Moreover, do we really see a body that always and everywhere moves upward? Do we really see that the element fire always rises, and that it never sinks, even a little, in the direction of the earth? Let me suggest, then, that the true meaning of this last assertion about upward and downward motion is something different from what Aristotle's surface argument has led us to expect. And he does not even claim, in fact, to have shown that there is one body that always and everywhere moves upward and another that always and everywhere moves downward; his precise words, rather, are that there is both upward and downward motion everywhere and always. And it is manifestly true that the downward or upward motion of one body is everywhere and always accompanied by motion in the opposite direction on the part of the body that it displaces.[46] More importantly, as I have argued, given the absence of a simply light element that is wholly lacking in weight, the downward motion of one body is everywhere and always responsible for any natural upward motion on the part of another. And by referring here to the fact that upward motion is always *accompanied* by downward motion, Aristotle would be pointing in the general direction, at least, of this true account of it.

The Doctrine of Weight and Lightness   133

In the continuation of his argument, Aristotle turns away from the simply light to a consideration involving the intermediate bodies. If all the elements, he says, were made up of a single, heavy material, some of the intermediate bodies will move downward more quickly than earth, since there will be more parts of that material in a large amount of air, (for instance, than in a small amount of earth). Yet not even a single part of air, he objects, is ever observed to move downward. Now what he apparently means by this argument is that air would sink to the bottom of earth if it moved downward more quickly and that a sufficiently large amount of air would have to move downward more quickly than a small amount of earth—by virtue of its being heavier—if there were a single, common material of all bodies. Here again, however, Aristotle could not have been satisfied with the argument that he presents. For he will soon go on to say that shavings, as well as other small earthy and dustlike bodies, can be seen to float in air;[47] and whenever one of these bodies moves upward, some of the surrounding air must move downward to fill its place. To be sure, there may be special reasons, as Aristotle will himself suggest, for this phenomenon; and it is true that even a small amount of earth will normally sink to the bottom of air (or water). Thus, one might try to defend his argument as saying that if there were only a single, heavy material, some amount of air (or water) would have to move downward more quickly than some amount of earth, and thus sink beneath it, even in the absence of special causes, whereas in fact it never does. But even on this interpretation, Aristotle's argument is unconvincing, as we can see by contrasting it with his own definition of the relatively light.

What Aristotle has said, we recall, is that one body with weight (such as air, for instance) is lighter than another body if the speed of its downward motion is less than that of *an equal volume* of the other. This definition led to the question, which so far we have left unanswered, of why he limited this comparison to the speeds of bodies that are equal in volume.[48] But now I am ready to propose an answer to this question. To do this, let me first return to my own assertion about the cause of upward motion. For in claiming that the upward pressure from a surrounding fluid overpowers the downward tendency of any immersed body whose weight is less than that of the fluid itself, I did not give a precise enough statement of this principle. More precisely, the upward pressure from the fluid (whether it be a liquid or a gas) overpowers the downward tendency of the immersed body, thus forcing it to the surface, when the weight of that body is less than that of *an equal volume* of the fluid. To see that this is so, let us imagine, for instance, an immersed cube whose weight is less than that of an equal volume of the surrounding fluid; the total weight pressing down at the base of this cube will be that much less than the weight pressing down on any equal and adjacent surface area of the fluid itself; and as a result of this disequilibrium of pressure, some of the fluid will force its way in underneath the cube, thus pushing it upward.[49] Now of Aristotle's four elements, or at least of the three that we know from direct experience, only one of them, earth, is not a fluid.[50] And I claim that he knew enough about fluids to know that it is only the weights—or the downward speeds, as he presents it—of *equal volumes* of his elements that determine which of them will rise to the surface of the others. I also claim that he knew, therefore, that even if no amount of earth were ever observed to rise in air, this would not mean that a large amount of air cannot be heavier than a small amount of earth, and hence it would not prove that air and earth cannot be made up of a single, heavy material. And his definition of the relatively light—which seemed especially puzzling in the light of this very argument against there being a single, heavy material of bodies—is in fact his way of indicating his awareness that the argument is faulty.

Having argued, or appeared to argue, that there cannot be a single material of all the elements, Aristotle then restates his contention that even a pair of contraries, such as the full and the void, do not suffice to account for their various motions. In particular, he denies that the doctrine of the full and the void can explain why the lighter of the two intermediate elements always rises to the surface of the heavier one. For if air is lighter than water by having more of the upward moving fire, while water has more of the downward moving earth, there will still, he says, be some (large) amount of water that contains more fire than a little air does. Also, he adds, much air will contain more earth than a little water does, so that some amount of air would have to move downward more quickly than a little water. He objects, however, that this is never observed to happen.[51] Now I deny, of course, that Aristotle really accepted the implicit premise of this argument, that an amount of air weighing more than some other amount of water must move downward more quickly (and hence sink to the bottom of it). But the more obvious difficulty with the argument is that Aristotle draws the conclusion from only the second of its two explicit premises, as if he had never stated the first one. For in saying that some amount of air would have to move downward more quickly than water (if they were both made up of earth and fire), he is relying only on the premise that relates the amounts of earth in these two elements. And yet if he had also spelled out the consequences of the differing amounts of fire, he would have had to acknowledge that an amount of air large enough to contain more earth than does a small amount of water would have a still greater preponderance of fire. And if both earth and fire were intrinsically responsible for motion, then this large amount of air would not be expected to sink beneath the water.[52] Similarly, an amount of water large enough to contain more fire than does a small amount of air would have a still greater preponderance of earth, so that it would not be expected to rise above the air. In other words, the doctrine of the full and the void would seem to be compatible with the observed relationship of water and air as long as one assumes that it is the *difference* between full and void (or between earth and fire) that determines the weight or lightness of a body. And by proposing that fire and earth are each responsible for a kind of motion, but then drawing a conclusion as if only earth were directly involved, Aristotle again raises the question of why he has failed to mention this subtractive hypothesis.

When I discussed Aristotle's earlier failure, in his more extended critique of the doctrine of an interstitial void, to mention this subtractive interpretation of it, I said that this interpretation could not explain why even a small amount of fire is lighter than any amount of air (assuming that air is predominantly void), nor why even a small amount of earth is heavier than any amount of water (assuming that water is predominantly solid).[53] Now, however, we are ready to see that this apparent weakness could not have been the truest reason for his failure to mention it. And we can also now see that the apparently successful explanation that this hypothesis offers for the relationship of water and air is itself a mere illusion or, rather, that it is an illusory solution to an illusory problem. For it accounts for the fact that air always rises in water without challenging the false premise on which Aristotle has relied in his explicit argument against the doctrine of the full and the void, namely, the premise that air would not behave in this way unless any amount of it, however large, were lighter than any amount of water.[54] More generally, this subtractive hypothesis remains caught up in the attempt to justify the false assumption that no amount of a lighter element can be heavier than any amount of a heavier one. And by tempting us to imagine that this interpretation might be able to rescue the doctrine of the full and the void from his argument against it, but then pointedly failing to take advantage of it himself, Aristotle provides us with a still further indication that he was aware of the falsity of that assumption.[55] Though he continues to argue as if he thought that no amount of a lighter element could be heavier than any amount of a heavier one, he thus corroborates the suspicion that this was not his genuine view and that he knew, rather, that the separation of the elements into four distinct places depends only on the differences in the weights of equal volumes.

I have argued that Aristotle's true view of the sublunar bodies is that they all have weight, and that the lighter ones among them, including even the lightest one, are merely less heavy than the others. He understood, as I claim, that the natural upward motion of these lighter bodies results from the greater downward propensity of an equal volume of the body that surrounds them (or that they surround). Now if this is true, the cause of a body's motion to its own place can not be—at least not in the case of a lighter body—so simple as Aristotle has presented it as being. Though he has claimed that external causes play no direct role in such motion, but are involved merely in the body's generation and in the removal of hindrances to its own motion, this can not be his true position. Rather, as I suggested earlier, he understood that the external pressure from the greater weight of the surrounding body plays as direct a role in accounting for natural upward motion as does any internal principle in the lighter body itself.[56] It follows from this, moreover, that the sharp distinction that Aristotle appears to draw between natural motion and violent motion, a distinction based on whether the principle of the motion is in the being itself or in something different, can also not have been his last word.[57] For he knew that the lighter bodies, at any rate, would not have their own natural propensities if they did not coexist in a world with other, heavier bodies that can force them upward. And an indication that he was aware of the provisional character of his sharp distinction between motion with an internal principle and that with an external one can be seen in the following important statement: "the order proper to the perceptible beings is [their] nature" [ἡ γὰρ τάξις ἡ οἰκεία τῶν αἰσθητῶν φύσις ἐστίν, 301a5–6]. Aristotle says in this way that the nature of a being is inseparable from the perceptible order to which it belongs, and hence from the natures of those other beings that regularly act upon it and are acted upon in turn. And this implies, I contend, that it is only in the often violent relationship among bodies that each of them, and most especially each lighter one, has its natural motion to its own place.[58]

According to Aristotle's explicit presentation, the natural order of the four elements consists of places in which each of them is at home, as it were, by virtue of having fully attained its proper form as a heavy or light body. What he teaches indirectly, however, is that this order is more complex: the lighter bodies, in particular, are forced upward to their own places, and they are so far from simply attaining their proper form as light bodies there that they have neither their upward propensity nor any peaceful rest, but rather a propensity downward that is normally thwarted by the presence of the heavier bodies beneath them.[59] And as we reflect on this situation, we are led to wonder what, if anything, guarantees the permanence of such a natural order. For Aristotle has said that what is violent and contrary to nature can not be eternal, and hence can not be a part of the eternal order of the world.[60] How, then, if the lighter bodies rest naturally in their own places only as a result of violent opposition to their downward propensities, can he still maintain that their ordered arrangement will last forever? This question is especially pertinent, moreover, since Aristotle has never completed the argument for the eternity of the world that he began in book one of *On the Heaven*. Though he has argued that the circular motion of the heavenly body is eternal, his promised argument that the world as a whole is also eternal turned instead into a series of arguments that nothing that has come into being can be eternal and that nothing that has not come into being can be destroyed. He did suggest that these new arguments would also make it clear whether there can be generation and destruction of infinitely many different worlds or whether this one must exist, or at least recur, eternally; but there is nothing in the arguments themselves that convincingly supports either of these latter alternatives.[61] Let me suggest, then, in the light of all this, that one of the main results of the whole discussion of weight and lightness, and indeed one of its main purposes from the beginning, is to confirm a suspicion that Aristotle was aware of the dubiousness of his own doctrine of the eternity of our world.[62]

Let me now turn my attention briefly to the last chapter of *On the Heaven*, which appears as a kind of appendix to the discussion of weight and lightness. Aristotle raises the question there of why flat bodies composed of iron and lead float in water, whereas others, which are smaller and less heavy, move downward if they are elongated or round. He also asks why some things float on account of their smallness, as for instance shavings and other earthy and dustlike bodies in the air. Concerning all these matters, he replies, it is incorrect to believe (νομίζειν, 313a21) that the cause is the one put forward by Democritus. For Democritus, he continues, says that there are warm particles that move upward from water, and that they support the weight of flat bodies, whereas narrow bodies, which are opposed by only a few of these particles, fall through. Aristotle objects, however, that on this view flat bodies should float still more easily in air than they do in water, since air, which is a warmer element, should have more of these upward moving particles. Now Democritus, it turns out, had raised this objection against his own theory, and he had tried to resolve it by proposing that the upward moving particles do not combine into a unity in air, as they do in water. Yet Aristotle is not persuaded by this attempt on Democritus's part to save his theory. As an alternative, he offers his own explanation of the phenomena in question. He begins his explanation by saying that since some continuous bodies are more easily divided than others, and since some are likewise more capable than others of causing

The Doctrine of Weight and Lightness 139

division, one must believe (νομιστέον, 313b8) that these are (the) causes. He continues with the claim that what is easily adaptable to the shape of its surroundings (i.e., a fluid) is easily divisible, and more so the more adaptable in shape it is. He also claims that a smaller amount of every kind of body is more easily divisible and split apart than a larger one.[63] And on the basis of these premises, he concludes that flat bodies, because they cover a large amount of fluid (and also because they are less capable of causing division), remain floating, whereas oppositely shaped bodies sink downward, both because they cover only a little of the fluid and because they divide it easily. This account, moreover, allows Aristotle to explain why bodies have a greater tendency (as far as their shape is concerned) to sink in air than they do in water, since air is of course the more easily divisible medium.

Now in the light of Aristotle's own explanation of why flat bodies float more easily than others, one can see why he rejected the explanation put forward by Democritus. For Democritus's account relied on the hypothesis that there are warm particles always moving upward from within any body of water. And to defend this hypothesis against an objection that he had raised himself, he proposed the further hypothesis that these hypothetical particles also exist in air but behave differently in air than they do in water. And yet a doctrine that thus relies on immanifest hypotheses, however plausible it might be, can never be known to be true. The only evidence, after all, for Democritus's claims that there are such particles as he supposes within water and within air is the success of these hypotheses in explaining observed phenomena; but even if the explanations are plausible, the possibility can never be ruled out that some radically different hypothesis might account for the same phenomena as well or better. Accordingly, Aristotle resists the temptation to speculate about such causes, and he attempts instead to account for these anomalous instances of floating in terms of the manifest properties of the beings that we actually observe. More generally, he tries to explain those features of the given world that require, or allow for, explanation in terms of other, more primary features of the given world itself. And this limitation on his part is not a sign of naivete, but rather of his keen awareness of the requirements of genuine intelligibility.[64]

To return, however, to the particular questions that Aristotle has raised in this chapter, we note that he does not explicitly attempt to resolve the question of why small, earthy bodies float in air. And his explanation of why flat bodies, despite their being heavier than water, can float in it is not applicable, or at least not clearly so, to this other question. For the specks of dust that we see floating in air are not typically flat in shape. And if, moreover, as Aristotle claims, a small amount of a bodily medium is more easily divisible than a larger one, it could well seem that these small, earthy bodies should sink in air, rather than float. On the other hand, we know that specks of dust float in sunbeams, where the air is warmer than it is in shade, and we also know that warm air rises. Thus, it appears that Democritus's hypothesis of upward moving warm particles within the air might indeed be at least part of the reason for these instances of floating. And in this connection, it is worth reminding ourselves of the exact words with which Aristotle first rejected that hypothesis: it is incorrect to "believe," he said, regarding "all" these matters that have been brought up, that the cause is the one put forward by Democritus. But doesn't this formulation leave open the possibility that Democritus's proposal might be the correct account of *some* of the phenomena in question? To be sure, the force of the word *all* in Aristotle's remark is ambiguous. But precisely this ambiguity points to a still further one, and suggests that what Aristotle means, above all, is that we should not "believe" that Democritus has given the true cause of any of these matters, even though he might in fact have been correct about some of them. In other words, this doubly ambiguous remark supports my suggestion that the root of Aristotle's disagreement with Democritus is a difference as to what *kind* of explanation is appropriate to natural science. For Democritus's proposal, at least insofar as it speaks of air as being made up of discrete particles, is an immanifest hypothesis, which as such could never be known to be true. And on these grounds alone, Aristotle would deny that it can be an adequate explanation of anything. Nevertheless, by ending his discussion of weight and lightness with the question left unresolved of why small, heavy bodies float in air, he may be directing our attention to a whole class of phenomena that also cannot be adequately understood in terms of manifest causes.[65] It would not be inappropriate, at any rate, for him to conclude *On the Heaven* by calling our attention in this way to the limits of genuine science.

NOTES

1. *On the Heaven* 276a18–b11, 278b9–21, 300b16–301a11, 307b18–24; cf. *On Coming into Being and Perishing* 329b7–330a29.
2. *On the Heaven* 269a2–18, 269b18–270a12.
3. *On the Heaven* 302b5–9, 303b4–8, 304b11–22.
4. *On the Heaven* 308a7–13, 308a34–b2; 277a33–b9, 310a7–15.
5. Cf. Simplicius, *In Aristotelis De Caelo Commentaria*, in *Commentaria in Aristotelem Graeca*, vol. 7, ed. I. L. Heiberg (Berlin, 1894), 674–75.
6. *On the Heaven* 308a7–33; compare 308a14–17 with 309a22–23; cf. also 270b1–11, 284a2–18.
7. *On the Heaven* 309b12–15; cf. 301a26–b13.
8. Cf. Simplicius, *In Aristotelis De Caelo Commentaria*, 678.25–34.
9. *On the Heaven* 308b21–28, 311a22–29, 312b32–313a13.
10. *Timaeus* 56a6–b3, 58d8–e2. The notion of weight and lightness, however, that is suggested in these passages, and that Aristotle criticizes, is only partially consistent with the one presented in the thematic discussion at 62c3–63e8.
11. Cf. *Meteorology* 340b21–27, 341b13–24. See also *On Coming into Being and Perishing* 330b22–30.
12. *On the Heaven* 308b30–309a11.
13. *On the Heaven* 309a19–b16.
14. The pertinence of this question is highlighted by the fact that the word ἀνάλογον (or ἀναλογία), which Aristotle uses here to mean sameness of ratios, is one that he also uses to mean sameness of differences, i.e., an arithmetical, rather than a geometrical, proportion. (*On the Heaven* 309b8–12; for the distinction between the two senses of the word ἀναλογία, see *Nicomachean Ethics* 1106a33–36 and 1131a30–1132a7.) It is also noteworthy that earlier in the very discussion we are now considering, Aristotle uses the word ἀναλογία in an apparently looser sense, which seems, however, to suggest an arithmetical at least as much as a geometrical proportion. (*On the Heaven* 309a14; compare the footnote by W. K. C. Guthrie on pp. 336–37 of his Loeb Classical Library translation of *On the Heaven* [1939].)
15. *On the Heaven* 309b34–310a3; cf. 311a22–29. For other interpretations of this passage, see Simplicius, *In Aristotelis De Caelo Commentaria*, 691.29–692.11.

16. *On the Heaven* 309b29–33, 310a3–11; cf. Simplicius, *In Aristotelis De Caelo Commentaria*, 690.22–691.2.
17. *On the Heaven* 310a16–b24, esp. 310a31–b1.
18. *On the Heaven* 311a1–6; cf. *Physics* 255a30–b13.
19. *On the Heaven* 310b26–29; cf. *Metaphysics* 1032b21–26.
20. *On the Heaven* 311a8–9.
21. *On the Heaven* 307b31–33; cf. 297a8–9 and b3–7.
22. Cf. *Physics* 212a6–28.
23. Cf. *On Coming into Being and Perishing* 331a12–b2.
24. Cf. *Physics* 212b33–213a6; *On Coming into Being and Perishing* 335a14–22.
25. Cf. *Physics* 211b10–14.
26. *On the Heaven* 310a31–b1, 310b16ff.
27. *On the Heaven* 311a12–16; cf. 310a16–20, 311b1.
28. *On the Heaven* 269b23–29; cf. *On Coming into Being and Perishing* 317b5–7.
29. *On the Heaven* 311a29–b13.
30. Simplicius (*In Aristotelis De Caelo Commentaria*, 710.14–711.25) claims to have performed this experiment and to have found no difference between the weights of the inflated and the uninflated bladders. He also refers to others, including Ptolemy, who reported that the inflated bladder was even slightly lighter than the uninflated one.
31. For a fuller discussion see, for instance, Archimedes, *On Floating Bodies* (ed. T. L. Heath [New York: Dover, 1912]) or Pascal, *A Treatise on the Equilibrium of Liquids*.
32. *On the Heaven* 311b13–24; cf. 312a3–5.
33. *On the Heaven* 311b21–27.
34. See, for instance, *Physics* 216a27–29.
35. *On the Heaven* 311b29–312a6.
36. *On the Heaven* 312b2–19.
37. *On the Heaven* 312a17–19, 30–33; contrast 309b29–34, 310a3–15.
38. Aristotle does not think that air would rise on its own to fill the place vacated by fire; for if he did, he would not have said that it can move upward only by violence. Rather, just as he treats its downward motion, whenever the water underneath it is removed, as a sign that it has weight in its own place, he would have treated such upward motion as a sign that it also had lightness in its own place.
39. Averroes, however, did try to resolve this question of why the cooling air should draw water upward into the vessel by appealing, among other causes, to the principle that nature seeks to avoid a vacuum, and he reports that Alexander of Aphrodisias had done so as well.

(Averroes, *Aristotelis De Caelo . . . cum Averrois Cordubensis variis in eosdem commentariis*, in *Aristotelis Opera cum Averrois Commentariis*, vol. 5 [Venice: Junctas, 1562–74; Frankfurt: Minerva, 1962], p. 265[H]–66[C].) On the other hand, Aristotle's own sympathetic account of the explanation given by Democritus and others as to why water remains suspended in a *klepsydra*—namely, that the air underneath it has no place to move—shows that he was quite capable of understanding such phenomena without appealing to any so-called abhorrence of a vacuum. See *On the Heaven* 294b13–30, esp. b25–28, and compare *Physics* 213a22–27. (A *klepsydra* was a container with a large opening in its narrow top and with many small ones in its wide base—somewhat similar in appearance to an upside-down funnel. As long as the opening in the top is sealed, water remains suspended in the *klepsydra*, and it does not escape through the holes in the base.)

40. The air in the vessel pushes downward, as it cools, with less pressure than that from the atmospheric air on the water adjacent to the vessel's opening. Thus, the water within the vessel rises to a higher level than that of the surrounding water. Cf. Pascal, *Treatise on the Weight of the Mass of the Air*, chap. 2, sec. 6.

41. The claim that the four elements share a common material is a provisional one, which Aristotle will soon again deny (*On the Heaven* 312b19–21; cf. p. 132). Indeed, I have argued at length in the previous chapter that he in fact regarded the matter of each element as being inseparable from it (cf. pp. 82–86). But his temporary adoption here of a view which he rejects serves the purpose of pointing to his (concealed) agreement with his predecessors that the elements, if unimpeded, all tend to move downward.

Compare J. L. Stocks's translation of *De Caelo*—n. 3, at 312b19—in *The Works of Aristotle*, ed. W. D. Ross, vol. 2 (Oxford: Clarendon Press, 1930).

42. *On the Heaven* 277b12–24.

43. Aristotle also includes a brief argument that the intermediate body cannot be outside of the places of the other two in a way contrary to its nature, but this part of his argument is not relevant to us now.

44. Here, for instance, is the note to this passage from Guthrie's translation of *On the Heaven*: "The argument is expressed with more than usual awkwardness. τὸ ἐπιπολάζον would naturally mean τὸ πᾶσιν ἐπιπολάζον, i.e. fire (312a4 below), but must here refer to the intermediate body, which rises to and stays on the surface of earth. It must stay there, and not pass upward through fire to the outside, for of a weightless body (fire) and a body possessed of a certain weight (the intermediate body), the latter (τὸ βάρος ἔχον) will occupy the lower place. The rea-

son for this is that the place of the absolutely heavy (τὸ βαρύ) is the centre, and therefore the relatively heavy must be nearer the centre than that which has no weight at all. I believe, though it is perhaps doubtful, that A. is leaving his fifth body out of the account, and is supposing a Universe constructed only out of the four elements recognized by previous and contemporary thought . . . For if τὸ κύκλῳ φερόμενον = *aither* and τὸ ἐπιπολάζον = fire, it would seem that fire must be the 'body with weight' of the next sentence" (80–81). See also Simplicius, *In Aristotelis De Caelo Commentaria*, 272.6–273.21, esp. 272.35–273.8.

45. Contrast *On the Heaven* 312b19–21ff. with 312a30–33. See also n. 41.

46. See, again, *Physics* 216a27–29.

47. *On the Heaven* 313a19–21ff.

48. See pp. 116–17.

49. See p. 126 and n. 31. This account also explains, of course, why a fluid whose weight is less than that of an equal volume of an immersed body allows the body to descend, with a corresponding upward displacement of its own. For a more complete demonstration of this principle, see Archimedes, *On Floating Bodies*, and in particular, propositions 4 through 6. Note also that the larger the immersed body of a given kind, the greater the difference will be between its weight and that of an equal volume of the surrounding fluid, and hence the greater the force pushing it upward. Thus, despite Aristotle's earlier claim to the contrary (*On the Heaven* 277a33–b5), this understanding of buoyant force as the cause of lightness is consistent with the fact that a larger amount of an upward moving body rises more quickly than does a smaller one. (It is true that the larger of these two [equally dense] bodies offers proportionally more resistance to the greater force pushing it upward; but the other element of resistance, from the fluid medium—which resistance depends on the [upper] surface area [and on the velocity], rather than on the total volume, of the immersed body—does not increase by a proportional amount, at least not if the two bodies are roughly similar in shape. Thus, the larger body will rise more quickly, and attain a greater terminal velocity, than the smaller one.)

Aristotle also argues in that earlier passage that this explanation of lightness in terms of upward pressure from the surrounding body is inconsistent with the fact that fire, as well as earth, moves more quickly near the end of its motion. "For all things," he says, "move more slowly as they come to be further from that which forced them [to move]." Now to prove that Aristotle was not in earnest about this argument would require an extended discussion of his teaching regarding the cause of the continuance of violent motion after the original mover has ceased to be in contact with the moving body. Such a discussion goes beyond my purpose here. But we should note that a body being pushed upward by a surrounding fluid is constantly in contact with the fluid that presses in beneath it, so that it never comes to be, as Aristotle's argument suggests, removed from some primary agent of its motion. Thus, even if we cannot (on Aristotle's stated principles) account for any increase in the body's speed, we would not, at any rate, expect it to move more slowly as it rises (*On the Heaven* 277b5–7; cf. 277a27–29, 301b17–30; *Physics* 266b27–267a20, 215a14–17, 243a18–b2).

As for Aristotle's argument at *On the Heaven* 277b7–8, it must be contrasted with what he says at 312b2–19, an argument that applies to the element fire, if it exists, as well as to air and water.

50. The case of fire is difficult. Aristotle speaks of fire, like earth, as being dry; and he defines dryness as the tendency of a body to preserve its own shape, and not to be adaptable, as is the wet, to the shape of contiguous bodies (*On Coming into Being and Perishing* 329b30–330a1, 330b3–4; cf. 335a14–21). Accordingly, if earth were surrounded by fire, neither of them, it seems, would behave like a fluid. However, Aristotle also speaks of fire as the finest of the elements, or the one with the smallest parts, and he says that fineness is a quality akin to, and even derivative from, wetness or fluidity (*On Coming into Being and Perishing* 329b34–330a4, 332a20–22; cf. *Parts of Animals* 649b16). The fineness of fire, then, could account for it being easily split apart, and hence displaced, like a fluid, by the relatively heavier earth (cf. *On the Heaven* 313b8–10ff.).

51. *On the Heaven* 312b32–313a6.

52. Cf. Galileo, *Discourse on Bodies in Water*, trans. Thomas Salusbury, with introduction and notes by Stillman Drake (Urbana: University of Illinois Press, 1960), 69–70.

53. See p. 120.

54. The argument of this subtractive hypothesis, though applied to fire and earth, rather than air and water, is spelled out on p. 119. Note that in this latest passage of the text, as opposed to the earlier one, it is suggested that fire might consist of void, and nothing else (*On the Heaven* 313a1–2, 7–8). Could Aristotle be hinting in this way that he did not even think there is such an element as "fire"? Given his explicit characterization of the element fire as wholly lacking in weight, and in the light of our discussion about why the lightest element rises to the surface of all the others, this question should not seem incredible. See also p. 117 and n. 11, and (regarding the elements more generally) p. 115 and n. 3.

55. For another interpretation of this difficulty in the text, see Simplicius, *In Aristotelis De Caelo Commentaria*, 727.29–728.10.

146   AN APPROACH TO ARISTOTLE'S PHYSICS

56. See pp. 121–22. Since the external pressure results from the relative weights of (equal volumes of) the two bodies, a sharp distinction between internal and external principles is not even possible.

57. *On the Heaven* 301b17–22ff.; cf. 300a23.

58. The existence of an order encompassing the natures of the various beings does not entail by itself any particular relationship among the elements. Yet Aristotle offers a further hint, at the very end of *On the Heaven*, of his awareness that even natural motion involves an opposition between the inner tendencies of several bodies. For in the next to the last sentence of the work, in the course of explaining why flat bodies tend to fall more slowly than round or elongated ones, he acknowledges that all heavy bodies must *force* their way downward (βιάσεται κάτω, *On the Heaven* 313b20) against the tendency of the medium to resist separation. By thus ending the whole work with a reference to the violence that is present in natural downward motion, he helps to confirm our impression that he understood the still more important aspect of violence, or of external force, in the natural upward motion of the lighter bodies.

59. This fact that the lighter bodies exchange their upward propensity for a downward one as soon as they reach their own place may help to explain Aristotle's strange example of the transformation of nourishment, rather than the growth of the being that it nourishes, as an analogy for a body's motion to its own place. Cf. p. 122 and n. 20.

60. *On the Heaven* 296a24–34. This claim refers specifically to violent and unnatural motion, rather than rest, since Aristotle is arguing against the view that the earth moves (with a motion different from its natural motion toward the center). Clearly, however, the claim also applies with equal force against the supposition that there can be permanence of violent and unnatural rest.

61. *On the Heaven* 279b4–284a2, esp. 280a23–30. The arguments referred to in the text do indeed assert that everything that is capable of both being and not being has a limited time in which it can *not* be, as well as a limited time in which it can be (*On the Heaven* 281a28–33, 282b10–14; cf. 283a12ff.). And this claim does seem to rule out at least the permanent annihilation of our world, even on the hypothesis that it is not simply eternal (cf. Simplicius, *In Aristotelis De Caelo Commentaria*, 310.26–311.21). But Aristotle offers no convincing argument in support of this radical claim, a claim that would also rule out the possibility that even individuals can be strictly mortal (cf. Simplicius, *In Aristotelis De Caelo Commentaria*, 343.1–21ff.).

62. Cf. *Topics* 104b12–17; *On the Movement of Animals* 699b12–31; see also Maimonides, *The Guide of the Perplexed*, trans. S. Pines (Chicago:

University of Chicago Press, 1963), pt. 2, chap. 15, 289–93.

63. Cf. *On the Heaven* 305a4–14; *On Coming into Being and Perishing* 328a23–28ff.

64. Cf. *On the Heaven* 306a9–11; Simplicius, *In Aristotelis De Caelo Commentaria*, 730.17–27. See also Albert Einstein and Leopold Infeld, *The Evolution of Physics* (New York: Touchstone Books, 1966), 31 and pp. 33–34 above.

65. In this regard, it is worth noting that Aristotle does not claim to know with certainty even why flat, heavy bodies float in water. For by introducing his proposed explanation with the expression "one must believe [νομιστέον] [that these are the causes]" (*On the Heaven* 313b8), he acknowledges some doubt as to whether he is right. Consider Galileo's powerful objection that if Aristotle's account were correct, and if resistance of a fluid to being divided were the reason that flat, heavy bodies can float in water, then those heavy bodies that float on the surface of water should also remain suspended underneath it, whereas in fact they sink (Galileo, *Discourse on Bodies in Water*, 40, 43–44).

CHAPTER 6

On Aristotle's Manner of Writing

Let me now turn thematically to the inevitable question of why Aristotle would go to such trouble to misrepresent his views about the topics I have been considering. Now as I suggested in my introduction, I think that his primary motive was concern for the reputation of natural science, whose very survival was in danger from powerful political forces.[1] It is hard for us today to appreciate that danger, now that freedom of speech is generally acknowledged to be a right and that modern natural science, with its accompanying technology, has become so indispensable a part of our civilization. But in ancient Athens, students of nature were suspected of atheism, which was a serious crime, and that suspicion played a major role in a number of persecutions, most notably the trial and death of Socrates.[2] Confronted, then, with this precarious situation for the study of nature, Aristotle undertook the project of trying to make it a permissible, and even a respectable, activity.[3] For this reason he deliberately exaggerated the degree to which natural science could support certain popular beliefs, such as the belief that our visible world will continue to exist forever or the belief that at least some of the changes that we see occurring around us are directed by higher causes for good ends.[4] By such rhetorical accommodations, he represented natural science as being more friendly than it can really be to religious or quasi-religious hopes. He was, moreover, quite successful in winning respectability for the study of nature. One measure of his success is the tradition of commentary on his works that extended even through the darkest ages for nearly two millennia after his death. We largely owe to this tradition the little that still survives to us from the writings of his predecessors in natural philosophy. And it is at least an arguable contention that Galileo and his successors, whose criticism of Aristotelian doctrines played such an important role in the birth of modern science, would never have had a tradition of science to nourish their own thinking had it not been for Aristotle's rhetorical success.[5]

However, fear of persecution and concern for the political respectability of natural science can not be the only reasons for Aristotle's willingness to misrepresent his views about nature. For those concerns by themselves do not account for the full character of his surface teaching. To begin, then, to see the further reasons for his deceptive manner of writing, let us briefly contrast what he says openly regarding the topics I have been discussing with the views that I argue he really held. He teaches openly that our visible world has always existed and will always exist; that one of its principles is a substrate that persists throughout all change; that there are also eternal and changeless forms, which by their action upon this substrate give rise to the natural beings; that the development of a natural being is set in motion for the sake of the good toward which it tends; that the manifest or perceptible character of the world, as for instance the continuity of its bodies and of their motions, is also its most intimate character, even beyond the range of any possible perception; that the up and the down, as they appear to us on earth, are also features of the world in itself, which characterize the places in which light and heavy bodies fulfill their natures; and, more generally, that there are permanent and proper places for the several elements, including earth, whose proper place is the absolutely fixed center of the world. If my arguments have been correct, however, he was well aware of the dubiousness of the claim that our visible world exists forever; he did not believe that any natural beings come into being from a persistent substrate (and all the less, from the action of eternal and changeless forms on such a substrate); he did not seriously claim that the end toward which a natural being tends to develop is in any sense anticipated by the moving cause or causes of that development; he did not believe that the manifest or perceptible character of the beings also belongs to them independently of our perception, but rather focused on it in the belief that the beings as we perceive them are what we properly mean by the beings themselves; in saying that the up and the down and the other differences of place are not just arbitrary designations, but genuine features of the whole itself, he meant by "the whole" the experienced whole, which exists as such only for human beings; and similarly, in saying that the earth remains unmoved at the center of the world, he meant that what we experience as the stability of the earth beneath us is part of the normal human perspective, within which natural beings are seen in their truest light.

What emerges from this contrast between Aristotle's surface teaching and his genuine views is that in the former he presented the natural world as being far more completely intelligible than he believed it was. For if the visible world is eternal, and if natural beings result from the action of eternal forms on a persistent substrate, then it can seem intelligible at least in principle why there are the kinds of beings there are. There arises, of course, the question of why the forms are the ones they are; but even this question can appear to be answered sufficiently, if not completely, if everything natural is brought about for the sake of the good.[6] Moreover, if natural motion is directed toward the good, then the tendency of living beings toward the attainment of their mature forms can appear to be explained with the kind of clarity with which we explain our own purposeful actions. Even the rising and falling of light and heavy bodies can appear to have the intelligibility of end-directed motion if in moving to their proper places they are becoming more completely the kinds of beings they are.[7] And finally, if we accept Aristotle's arguments for infinite divisibility, then at least to this extent, our knowledge need not be limited to the beings as we perceive them, but can encompass what lies beyond the perceptible realm. In Aristotle's serious view, however, natural beings do not arise from the action of eternal forms, and neither is there any other principle that can make intelligible—except conditionally, given that there is in fact an ordered world—why there are any of the kinds of beings that we know there are. As for the growth of these beings, he knew from experience that they typically develop in more or less the way their parents did, and he understood that their development is necessarily somehow coherent if they are to live; but he also knew that he could not understand the full necessity for this development or any purpose directing it.[8] Though he did have an explanation of why lighter bodies tend to rise, his account of this phenomenon relies on the more primary fact, which he did not really try to explain, that the medium in which they do so presses downward. And more generally, he regarded the task of natural science to be the articulation of the manifest character—understood as the truest being—of the given world, a world whose ultimate roots he did not think that this science could ever discover.

Now to understand why Aristotle presented what he knew to be such an exaggerated picture of the degree of intelligibility of the natural world, we must consider the implications of the limitedness of the achievement of what he regarded as genuine natural science. For his denial that natural science can finally explain the given world—and in particular his acknowledgment that it can not discover its ultimate roots—seems to leave him unable to exclude the alternative that this world might partly consist of, or otherwise owe its existence to, a mysterious and all-powerful god or gods. If there are such gods, as was suggested by Homer and Hesiod, among others, we can not rely on what reason and normal experience indicate as to the limits of what beings can do and of what can be done to them.[9] What I have called the manifest character of things could not be their truest being but at most their usual way of being, and the most important truth about them would have to include their capacity to undergo miraculous change. In other words, so long as this theological alternative is not ruled out, the very assumption that there is nature, or that beings must become and perish in accordance with fixed natures, remains questionable; and the pursuit of a science of nature remains a dubious and perhaps even a wholly misguided one. Aristotle must refute, then, the claims of this theology in order to vindicate the possibility of natural science. And the primary reason for his exaggerating the intelligibility of the world is to indicate what would have to be the case in order for natural science to be able to complete this task itself.

But the world is not intelligible in the way that Aristotle pretends, and so the question arises of what his ultimate response was to the challenge from those who deny that it is even a natural one. Now Aristotle says in the *Physics* that it is ridiculous to try to prove that there is nature, since it is clear that there are many natural beings. He says that those who try to prove this cannot distinguish between what is knowable in itself and what is not, so that their argument is about names, but they have no insight. It is clearly possible, he adds, to be deluded in this way, just as a man who has been blind from birth might make syllogisms about colors.[10] Thus, he likens the difference between those who know that there is nature and those who do not—in which latter class he includes not only those who deny its existence, but some who confidently assert it—to the difference between those with eyesight and those who are blind. Now by treating it as self-evident that nature indeed exists, Aristotle implicitly rejects the view that he seemed to endorse, in his discussion of the principles of the natural beings, that only an explanation of how their coming into being takes place could justify the claim that they are truly natural.[11] But how, then, even on the assumption that he is here speaking from genuine knowledge, can he hope to lead us his readers to a position from which we can acquire it ourselves? Now this question is too large for me even to attempt to resolve it here. But let me suggest, in conclusion, that the chief way in which he tries to lead us toward his own perspective on the world is through his *political* philosophy, which we find above all in the *Nicomachean Ethics* and the *Politics*. For in his political philosophy he tries to guide us toward an understanding of the superiority of the theoretical life or the life of science, in part by helping to remove the barriers—such as the belief that something can come from nothing—that obstruct our insight into the nature that makes it possible.[12] Aristotle also speaks explicitly of how this insight is to be approached at the beginning of the *Topics*, his work on dialectical reasoning. There he suggests that the task of clarifying the first principles of every science belongs uniquely or primarily, not to physics or even metaphysics, but to dialectics, and to its conversational scrutiny of accepted opinions.[13]

NOTES

1. See above, pp. 6–7.
2. See above, introduction, n. 23.
3. According to Plutarch, it was not Aristotle but Plato who succeeded in making natural philosophy respectable among the Greeks (*Life of Nicias* 23.2–4). But even granting that the first or decisive steps were taken by Plato, the fact remains that Plato's dialogues make light of the study of nature, at least as compared to the study of the changeless forms (e.g., *Phaedo* 95e8–100a3; *Timaeus* 27a3–6, 29b1–d3, 59c5–d2). Thus, it remained to Aristotle to present the study of nature as an explicitly serious part of a philosophic doctrine that could—and did—become widely respectable.

4. See above, pp. 22, 40, 101, 137–38. Regarding the claim that our visible world will continue to exist forever, compare Homer, *Iliad* Θ 18–27, and Hesiod, *Theogony* 116–28, with *On the Heaven* 270b1–16, 284a2–b5, 308a17–29; *Metaphysics* 1050b22–24.

5. Cf. A. N. Whitehead, *Science and the Modern World* (New York: Mentor Books, 1948), 6–8.

6. Compare *Metaphysics* 1075a11–25, *Politics* 1256b15–22, and *On the Generation of Animals* 731b18–732a1 with Plato, *Phaedo* 97b8–98b6 and *Timaeus* 29e4–31a1.

7. Cf. Whitehead, *Science*, 8–9.

8. Cf. Plato, *Phaedo* 96c1–97b7.

9. See the references to Homer and Hesiod in n. 4 above, as well as Homer, *Odyssey* κ 229–43 (but contrast 281–306); λ 601–4; and Hesiod, *Theogony* 950–55. Compare also chap. 1, p. 23.

10. *Physics* 193a3–9; cf. 184a16–21.

11. See above, pp. 22–24.

12. Consider the following suggestion of Alfarabi, which occurs in his extended account of the philosophy of Aristotle. According to Aristotle, says Alfarabi, some of the sciences that man desires "are firmer and some shakier than others. However," he continues, still speaking for Aristotle, "once [man] attains certainty about what he was investigating, this is the perfect science of what he wants to know and the end beyond which he can hope for no better assurance and reliability. This, then, is man's situation with regard to the *practical* sciences" (Alfarabi, *Philosophy of Plato and Aristotle*, trans. Muhsin Mahdi [Ithaca: Cornell University Press, 1962], 74, emphasis mine). To help understand Alfarabi's remark, see Christopher Bruell, "Strauss on Xenophon's Socrates," *Political Science Reviewer* 14 (1984): 263–318.

13. Contrast *Topics* 101a25–101b4 with *Metaphysics* 1004b17–26. Compare *Physics* 184b25–185a3, *Posterior Analytics* 77a26–35, *Metaphysics* 1005b35–1006a26, and *On Sophistical Refutations* 171b3–6.

INDEX

Ahrensdorf, P., 11n.22
Alexander of Aphrodisias, 27n.17, 30n.40, 73n.24, 108n.13, 112n.73, 142n.39
Alfarabi, 6, 10n.21, 154n.12
Anaxagoras, 25–26, 29nn.38, 39, 66, 103
Apostle, H., 75n.54
Aquinas, 26n.6, 50n.17, 109n.39, 110n.53, 111n.57, 112n.73, 113n.81
Archimedes, 142n.31, 144n.49
Averroes, 142n.39

Bostock, D., 74n.47
Bruell, C., 154n.12

Charlton, W., 26n.4, 28n.21, 29n.27, 49n.14, 108n.25
Chroust, A-H., 11n.22
Cooper, J., 50n.33
Cornford, F. M., 112n.75
Coughlin, R. G., 109n.39, 110n.53

Darwin, 5
Darwin(ian), 1, 36
Democritus, 73n.21, 138–40, 143n.39
Derenne, E., 10n.22
Descartes, 10n.20, 32, 49n.2

Eddington, Sir Arthur, 2, 8n.2
Einstein, 33, 49n.4, 147n.64

Furley, D., 50n.31

Galileo, 10n.16, 113n.77, 145n.52, 147n.65, 149

Goethe, 9n.5
Guthrie, W. K. C., 141n.14, 143n.44

Heidegger, 10n.13
Heisenberg, 8n.1
Hesiod, 23, 29n.33, 79, 152, 154nn.4, 9
Hobbes, 11n.22
Homer, 152, 154nn.4, 9
Husserl, 8n.3, 9n.4
Hussey, E., 109nn.30, 31, 110n.51, 111n.64

Infeld, L., 33, 49n.4, 147n.64

Kahn, C., 50n.17
King, H. R., 109n.32
Klein, J., 8n.3
Kuhn, T., 10n.13

Leucippus, 73n.21

Maimonides, 51n.39, 146n.62
Melissus, 62
Monod, J., 31–32, 49n.1

Newton, 32–33, 110n.44
Nussbaum, M., 9n.6

Pascal, 142n.31, 143n.40
Philoponus, 109nn.32, 40, 111n.64, 113n.81
Plato, 5, 7, 10n.18, 11n.23, 27n.14, 49n.10, 81, 107n.10, 117, 153n.3, 154nn.6, 8
Plutarch, 10n.20, 153n.3

Ross, W. D., 26n.4, 74n.44, 110n.51, 112n.75, 113n.81

Sachs, J., 110n.45
Simplicius, 5, 10n.20, 26n.4, 27nn.12, 17, 72n.13, 73 nn.22, 24, 25, 33, 74n.44, 75n.68, 99, 107n.9, 108n.13, 109nn.31, 32, 40, 111n.62, 112nn.68, 71, 73, 113nn.81, 86, 90, 141nn.5, 8, 15, 142n.30, 144n.44, 145n.55, 146n.61, 147n.64
Socrates, 6–7, 149
Sorabji, R., 110n.51, 113n.83
Stocks, J. L., 143n.41
Strauss, L., 11n.24

Themistius, 5, 10n.19, 27n.17, 62, 72n.14, 73,n.33, 74n.45, 100, 111n.64, 112n.73, 113n.81

Virgil, 75n.69

Whitehead, 154nn.5, 7
Wieland, W., 9n.6, 10n.12, 50n.22, 75n.63

Zeno, 53–54, 66, 70, 80, 104

www.ingramcontent.com/pod-product-compliance
Lightning Source LLC
Chambersburg PA
CBHW021759230426
43669CB00006B/131